Cro-Magnon

Also by Brian Fagan

"Where We Saw a Whale"

The Great Warming

Fish on Friday

From Stonehenge to Samarkand (editor)

Chaco Canyon

Before California

The Long Summer

The Little Ice Age

Egypt of the Pharaohs

Floods, Famines, and Emperors

Into the Unknown

From Black Land to Fifth Sun

Eyewitness to Discovery (editor)

The Oxford Companion to Archaeology (editor)

Time Detectives

Kingdoms of Jade, Kingdoms of Gold

Journey from Eden

Ancient North America

The Great Journey

The Adventure of Archaeology

The Aztecs

Clash of Cultures

Return to Babylon

Quest for the Past

Elusive Treasure

The Rape of the Nile

Cro-Magnon

HOW THE ICE AGE GAVE BIRTH TO THE FIRST MODERN HUMANS

Brian Fagan

BLOOMSBURY PRESS
New York   Berlin   London   Sydney

Published by Bloomsbury Press, New York

All papers used by Bloomsbury Press are natural, recyclable products made from
wood grown in well-managed forests. The manufacturing processes conform to
the environmental regulations of the country of origin.

LIBRARY OF CONGRESS CATALOGING-IN-PUBLICATION DATA

Fagan, Brian M.
Cro-Magnon : how the Ice Age gave birth to the first modern
humans / Brian Fagan. —1st U.S. ed.
p.   cm.
Includes bibliographical references and index.
ISBN-13: 978-1-59691-582-4 (alk. paper hardcover)
ISBN-10: 1-59691-582-X (alk. paper hardcover)
1. Cro-Magnons. 2. Human evolution. 3. Glacial epoch.
4. Neanderthals. 5. Prehistoric peoples. I. Title.
GN286.3.F34 2010
569.9'8—dc22
2009025242

First published by Bloomsbury Press in 2010
This paperback edition published in 2011

Paperback ISBN: 978-1-60819-405-6

1 3 5 7 9 10 8 6 4 2

Typeset by Westchester Book Group
Printed in the United States of America by Quad/Graphics, Fairfield, Pennsylvania

To

Francis and Maisie Pryor

Archaeologists, gardeners, and sheep farmers,

with affection and respect and with thanks for many good laughs.

After all, they have turtles named after them . . .

A sudden intense winter, that was also to last for ages, fell upon our globe.

Louis Agassiz, *Geological Sketches* (1866)

Contents

Preface to the Paperback Edition

NEANDERTHALS DESTROYED BY GREAT ERUPTION, screamed the headlines. WE HAD SEX WITH NEANDERTHALS! announced another some months earlier. NEANDERTHALS WERE NOT STUPID, INDICATES NEW EVIDENCE.

You'd hardly think that our prehistoric predecessors could fall victim to celebrity gossip, but in a way, that's what has happened with some recent science news related to Neanderthals. While our knowledge of the Cro-Magnons themselves has changed little since I wrote this book, their Neanderthal predecessors continue to be the subject of aggressive research, which sometimes prompts seemingly sensational revelations. Archeological findings about our archaic forebears, unlike news about the Cro-Magnons, has earned a place in the public imagination. Every find, however trivial, seems to gain a life of its own. Over the past year, we've learned that Neanderthals in Spain had freckles and that another Spanish archaeologist has unearthed a Neanderthal bedchamber, "a cozy bedroom with hearth and grass beds." Not that modern humans have escaped scrutiny as well. In South Africa, we learned of claims that the ancestors of *Homo sapiens* may have developed bows and arrows as early as sixty thousand years ago. Recent DNA researches have focused not only on the interbreeding of Neanderthal and *Homo sapiens* but have also developed more accurate dates for the emergence and spread of moderns out of tropical Africa after two hundred thousand years ago. A feast of discoveries, to be sure, but has there really been a revolution in our understanding of Neanderthals and Cro-Magnons? Once you separate the chaff of headline-seeking press releases and delve into the scientific literature, only a handful of finds have altered the picture, and those not nearly as much as has some have claimed.

At least we can now be certain that Neanderthals and moderns had sex, a hotly debated topic for generations. Population geneticist David Reich of Harvard University and his colleagues have analyzed large numbers of DNA fragments from Neanderthals dating to about thirty-eight thousand years ago from Croatia, Germany, Russia, and Spain. By using rigorous anti-contamination measures, they were able to piece together about 60 percent of the Neanderthal genome. Once their analysis was complete, the research team compared Neanderthal DNA to samples taken from living people in southern and western Africa, China, France, and Papua New Guinea. They found that Neanderthal DNA more closely matched that of the living Chinese, French, and Papua New Guinean individuals in their study. All of them had between a 1 and 4 percent Neanderthal contribution to their DNA. At some point, then, modern humans migrating out of Africa encountered Neanderthals, perhaps between forty thousand and eighty thousand years ago in the Middle East, long after they split off from archaic people at least two hundred thousand years ago. However, the interbreeding also occurred before *Homo sapiens* colonized Europe and more distant parts of Asia.

The postulated interbreeding matches what little we know about the human populations of the Middle East at the time, where small numbers of both Neanderthals and modern humans lived alongside one another. If the DNA evidence stands the test of scientific time, then they also interbred, perhaps sporadically, over a long period of time. Neanderthal DNA is about 99.7 percent identical to our DNA, but *Homo sapiens* evolved distinct genes that were related to cognitive functions, metabolism, and the development of some skull features, the collarbone, and rib cage. Biologist Robert Lahn and his colleagues at the University of Chicago have shown that a gene for brain size called microcephalin underwent a significant change about thirty-seven thousand years ago. The modified variant, or allele, of the microcephalin gene, which is present in about 70 percent of the world's population, appears to confer some fitness advantage to those who possess it. The microcephalin gene originated in archaic populations about 1.1 million years ago, among ancestors of the Neanderthals. Thus, occasional interbreeding between humans destined for extinction and related species may have contributed to the evolutionary success of *Homo sapiens*.

Neanderthals may have also interbred with moderns in Europe, where they survived alongside Cro-Magnons for a long time. But, as I have argued in this book, Neanderthal populations were thin on the ground almost everywhere, so any sexual contact must have been rare at best. Perhaps it resulted from very rare encounters between solitary or near-solitary individuals rather than bands, where commonly shared prejudices, especially among modern human groups, might have militated against any form of close contact, let alone sexual intercourse. Unfortunately, the current genetic data doesn't allow us to tie down the exact places where such contacts may have taken place. Thus, the long-held theories that had *Homo sapiens* replacing Neanderthals without interbreeding now appear to be obsolete. In the context of *Cro-Magnon*, however, occasional sexual contacts and interbreeding make little difference to the overall history. Cognitively far superior newcomers possessed such advantages over the indigenous Neanderthals that the latter passed inexorably into extinction.

DESPITE THE HEADLINES, few experts believe that the Neanderthals were stupid. They were expert hunters with an intimate knowledge of their environments; they had to be just to survive in sometimes arid and often savagely cold landscapes. But they weren't fully modern humans with all the cognitive potential and abilities of *Homo sapiens*. New, more refined chronologies place the appearance of modern humans in sub-Saharan Africa about two hundred thousand years ago.

Recently discovered artifacts from South Africa made about seventy-five thousand years ago may reflect enhanced human cognitive abilities. Hunters at Blombos Cave fashioned long stone spearheads. They heated the stone, then removed fine flakes with a bone tool, a method known as pressure flaking that only came into use by Cro-Magnons about twenty thousand years ago. At nearby Sibudu Cave, small geometric artifacts from sixty-four thousand years ago display the characteristic wear of sharp tips that were once mounted on the ends of light wooden shafts. Microscopic traces of blood and bone survive on the edges of the points, also of the plant resin used to mount them. Archaeologist Marlize Lombard, who carried out the detective work, believes that these African hunters were using bows at least twenty thousand years earlier than humans elsewhere in the prehistoric world. Such hunting weapons require careful planning and innovation to develop.

Volcanic eruptions loom large in the *Cro-Magnon* story, notably the epochal Mount Toba cataclysm of seventy-three thousand years ago and the Campanian eruption in Italy of about thirty-eight thousand years ago. Now other volcanic events have been added to the historical equation, identified from ash deposits in Mezmaiskaya cave in Turkmenistan dating to about forty thousand years ago. Artifacts made by Neanderthals lie like a sandwich between the ash. In geological terms, the two eruptions are almost contemporary with the first settlement of Europe by moderns. Inevitably, this has led to speculations about Neanderthal extinction in the face of suffocating volcanic ash. With sparse, widely separated Neanderthal and modern populations, both of which were highly mobile, it seems unlikely that volcanic events on these relatively small scales would have wiped out either archaic or modern groups between the Atlantic and the Ukraine. Besides, we know that Neanderthals survived in Spain until at least thirty thousand years ago, if not later.

OUR PERCEPTIONS OF the Cro-Magnons have changed dramatically in recent years, thanks to increasingly fine-grained excavation methods and fast moving advances in everything from genetics to paleoclimatology. The story told in these pages, based on the cumulative enterprise that is modern science, brings a late Ice Age people to vivid life for the first time. We now know that Cro-Magnons were far more than skilled artists and expert technologists. They were adept, highly adaptable people, who adjusted brilliantly and effortlessly to brutal climatic regimes. A new generation of research is pulling aside the dark curtain that separates us from the long-vanished, intensely symbolic world in which they lived. For the first time, too, we can begin to explore the complex interactions between these fully modern humans with their awesome cognitive abilities and their Neanderthal neighbors in ways that were unimaginable even a decade ago. These slowly changing relationships between archaic and modern people will help us better understand our own biological and cultural diversity in a rapidly changing world.

Preface to the First Edition

FOUR DOTS MOVE ALONG A RIVERBANK in a black and gray Ice Age landscape of forty thousand years ago, the only signs of life on a cold, late-autumn day. Dense morning mist swirls gently over the slow-moving water, stirring fitfully in an icy breeze. Pine trees crowd the riverbank, close to a large clearing where aurochs and bison paw through the snow for fodder. The fur-clad Cro-Magnon family moves slowly—a hunter with a handful of spears, his wife carrying a leather bag of dried meat, a son and a daughter. The five-year-old boy dashes to and fro brandishing a small spear. His older sister stays by her mother, also carrying a skin bag. A sudden gust lifts the clinging gloom on the far side of the stream. Suddenly, the boy shouts and points, then runs in terror to his mother. The children burst into tears and cling to her. A weathered, hirsute face with heavy brows stares out quietly from the undergrowth on the other bank. Expressionless, yet watchful, its Neanderthal owner stands motionless, seemingly oblivious to the cold. The father looks across, waves his spear, and shrugs. The face vanishes as silently as it appeared.

As light snow falls, the family resumes its journey, the father always watchful, eyes never still. During the climb to the rock shelter, he tells his children about their elusive, quiet neighbors, rarely seen and almost never encountered face-to-face. There were more of them in his father's and grandfather's day, when he saw them for the first time. Now sightings are unusual, especially in the cold months. They are people different from us, he explains. They do not speak like we do; we cannot understand them, but they never do us any harm. We just ignore them . . .

Cro-Magnons and Neanderthals: this most classic of historical confrontations, sometimes couched in terms of brutish savagery versus human sophistication, has fascinated archaeologists for generations. On the one side stand primordial humans, endowed with great strength and courage, possessed of the simplest of clothing and weaponry. We speculate that they were incapable of fully articulate speech and had relatively limited intellectual powers. On the other are the Cro-Magnons, the first anatomically modern Europeans, with fully modern brains and linguistic abilities, a penchant for innovation, and all the impressive cognitive skills of *Homo sapiens*. They harvested game large and small effortlessly with highly efficient weapons and enjoyed a complex, refined relationship with their environment, their prey, and the forces of the supernatural world. We know that the confrontation ended with the extinction of the Neanderthals, perhaps about thirty thousand years ago. But how it unfolded remains one of the most challenging and intriguing of all Ice Age mysteries.

The Neanderthals appeared on the academic stage with the discovery of the browridged skull of what seemed to be a primitive human in Germany's Neander Valley in 1856. Seven years later, Thomas Henry Huxley's brilliant study of the cranium in his *Man's Place in Nature* compared the Neanderthal fossil with the skulls of humankind's primate relatives, chimpanzees and gorillas. The thought of a human ancestry among the apes horrified many Victorians. Public opinion carved out a vast chasm between archaic humanity, epitomized by the Neander Valley skull, and the modern humans discovered in the Cro-Magnon rock shelter at Les Eyzies, in southwestern France, in 1868. The Neanderthals became primitive cave people armed with clubs, dragging their mates around by their long hair. Unfortunately, the stereotype persists to this day.

Cutting-edge science paints a very different portrait of the Neanderthals. They were strong, agile people who thrived in a harsh, often extremely cold Europe, from the shores of the Atlantic deep into Eurasia, from the edges of the steppe to warmer, drier environments in the Near East. Neanderthal hunters stalked large, dangerous animals like bison, then killed them with heavy thrusting spears. They didn't have the luxury of standing off at a distance and launching light spears at their prey. But, for all their strength and skill, they were no matches for the Cro-Magnon newcomers, who, science tells us, spread rapidly across Europe around forty-five thousand years ago. Their hunting territories were small; they were thin on the ground; the routine of their lives changed infinitesimally from one year to the next.

When they arrived in their new homeland, the Cro-Magnons were *us*, members of a species with a completely unprecedented relationship with the world around them. Every Cro-Magnon family, every band, was drenched in symbolism, expressed in numerous ways. Well before thirty thousand years ago, Cro-Magnons were creating engravings and paintings on the walls of caves and rock shelters. They crafted subtle and beautiful carvings on bone and antler and kept records by incising intricate notations on bone plaques. We know that they used bone flutes at least thirty-five thousand years ago, and if they did this, they surely sang and danced in deep caves by firelight on winter evenings and at summer gatherings. Cro-Magnons ornamented their bodies and buried their dead with elaborate grave goods for use in an afterlife. No one doubts that Cro-Magnon symbolic expression somehow reflects their notion of their place in the natural world. But their perceived relationship to nature was poles apart from our own—they were hunter-gatherers and lived in a world that was unimaginably different from today's Europe. And their perceptions of the world, of existence, were radically different from, and infinitely more sophisticated than, those of the Neanderthals.

Cro-Magnon briefly explores the ancestry of the Neanderthals and the world in which they lived, then tries to answer the question of questions: What did happen when Cro-Magnon confronted Neanderthal? Did the moderns slaughter the primordial humans on sight, or did they simply annex prime hunting territories and push their ancient occupants onto marginal lands, where they slowly perished? Or did the superior mental abilities, hunting weapons, and other artifacts of the Cro-Magnons give them the decisive advantage in an increasingly cold late Ice Age world? Do we know what kinds of contacts took place between Neanderthal and newcomer? Did the two populations intermarry occasionally, trade with one another, even borrow hunting methods, technologies, and ideas from each other?

The answers to these questions revolve as much around the Cro-Magnons as they do the Neanderthals. Despite a century and a half of increasingly sophisticated research, the first modern inhabitants of Europe remain a shadowy presence, defined more by their remarkable art traditions and thousands of stone artifacts than by the nature of their lives as hunters and foragers, defined by the Ice Age world in which they flourished. *Cro-Magnon* paints a portrait of these remarkable people fashioned on a far wider canvas than that of artifacts and cave paintings.

I DECIDED TO write this book in the galleries of the National Museum of Prehistory in Les Eyzies, the small village in France's Vézère Valley that prides itself on being the "capital of prehistory." The upper gallery is a quiet place nestled against the great cliff that houses the huge Cro-Magnon rock shelters that once flourished nearby. I gazed at the rows of flint, bone, and antler tools against one long wall, neatly laid out in series, each with its correct archaeological labels and subdivisions. The history of the Neanderthals and Cro-Magnons unfolded like an orderly ladder of artifacts, ever smaller, ever more refined over time. I stared, confused, despite having had formal training in these very tool kits many years ago. Minute variations in one scraper form compared with another; small chisels with different working edges; antler and bone points that once armed lethal spears: the display seemed endless. After a few minutes, I realized that the casual viewer would learn almost nothing about the anonymous makers of these museum-perfect objects beyond the fact that they were able to make artifacts of all kinds. Many questions remained unanswered. Who were the Cro-Magnons? Where did they come from? How did they survive the dramatic changes of the late Ice Age climate tens of thousands of years ago? And how did they behave toward the beetle-browed Neanderthals who were living along the Vézère River when they arrived? The museum displays commemorated a past peopled not by human beings but by artifacts. For all intents and purposes, a rich and vibrant history of some of our remote forebears was dead to all but a small handful of specialists.

Everyone has heard of the artistic glories of Lascaux and Altamira, Font-de-Gaume, and Grotte de Chauvet. Books on Cro-Magnon art of all kinds abound, many of them illustrated with magnificent color pictures of carved antlers, woolly rhinoceroses, aurochs, and Ice Age bison. The authors write of gifted artists, speculate about the motives for the engravings and paintings, sometimes imagine shamans with supernatural powers conducting ceremonies far from daylight. Beyond this, if the people of the period are mentioned at all, it is as big-game hunters pitting themselves against a formidable bestiary. Few of these volumes explore the most fascinating questions about the first modern Europeans—the complex dynamics of their societies, the ancient rhythms of their annual round. And few of them examine the most fundamental questions of ancestry and cognitive skills. Art defines the Cro-Magnons in the public eye when, in fact, it was an integral part of a much larger existence.

Cro-Magnon is a story of hunters and gatherers who lived a unique adventure, whose earliest ancestors almost became extinct in the face of a huge natural catastrophe over seventy thousand years ago. It is a tale of ordinary men and women going about the business of survival in unpredictable, often bitterly cold environments that required them to adapt constantly and opportunistically to short- and long-term climate changed. These people were like us in so many ways: they had the same powerful intelligence and imagination, the ability to innovate and improvise that is common to everyone now living on earth. But they dwelled in a very different world from ours, one where premodern people still lived the same way they had hunted and gathered for hundreds of thousands of years. The history of the Cro-Magnons is the story of a great journey that began over fifty thousand years ago in tropical Africa and continued after the end of the Ice Age some fifteen thousand years ago. Above all, it's a story of endless ingenuity and adaptability.

WHEN I WAS researching *Cro-Magnon*, I walked along the bank of the Vézère near Les Eyzies on a gray summer's day. The great cliffs with their rock shelters loomed high above, lapped by the deep green of meadow and thick woodland. The river itself ran brown and swift, swelled by the heavy rain of recent weeks. I imagined the same landscape eighteen thousand years ago—much of it treeless, covered with stunted grass and shrubs, a world alive not with bustling humans and their automobiles but with browsing reindeer and red deer with great horns, with chunky wild horses in small herds. There would have been black aurochs with lyre-shaped horns, perhaps arctic foxes in their brown summer fur feeding off a kill, perhaps a pride of lions resting under the trees. If you'd been patient enough, you'd have seen the occasional humans, too. But you would have known they weren't far away—informed by the smell of burning wood, trails of white smoke from rock-shelter hearths, the cries of children at play. Then I imagined this world changing rapidly, soon becoming one of forest and water meadow, devoid of reindeer and wild horses, much of the game lurking in the trees. I marveled at the ability of our forebears to adapt so readily to such dramatic environmental changes.

Few humans have ever lived in a world of such extreme climatic and environmental change. Years ago, I sailed a small yacht through the narrow channels of the Danish archipelago. The deeper water passages twisted and turned, marked by tall poles, nothing else. A gentle breeze from astern carried us through the sinuous defiles at little more than walking speed, which was just as well, as we grounded in the mud several times. I thought of Stone Age hunters fishing and fowling among the nearby reeds; some of them perhaps once camped on the then-dry ground now beneath our keel, in the midst of a dynamic landscape now buried by higher sea levels that changed from one month to the next. These were people without metals, with the simplest of canoes, and with fishing gear and weaponry created from the few suitable materials close to hand. The adaptability and ingenuity of *Homo sapiens* lay before my eyes and was a comforting thought when I contemplated the huge climatic and environmental challenges that lay ahead in the twenty-first century.

Thanks to multidisciplinary science, we now know a great deal more about late Ice Age climate than we did a generation ago. Much of the raw material for this narrative does indeed come from artifacts and food remains, from abandoned hunting camps and the stratified layers of caves and rock shelters. New generations of rock-art studies not only in western Europe but all over the world have added new perceptions about the meaning of Cro-Magnon art on artifacts and cave walls. However, compared with even twenty years ago, our knowledge of Europe's first moderns has changed beyond recognition thanks to technology and the now well-known revolution in paleoclimatology—the study of ancient climate. Another revolution, in molecular biology, has added mitochondrial DNA (passed down through the female line) and the Y chromosome (roughly the equivalent in men) to the researcher's armory. We now possess far more nuanced insights into Neanderthal and Cro-Magnon life, especially into the environments in which they lived.

Humans have always lived in unpredictable environments, in a state of flux from year to year. Until recently, we thought of the last glaciation of the Ice Age as a continual deep freeze that locked Europe into a refrigerator-like state for over one hundred thousand years, until about fifteen thousand years ago. Thanks to ice cores, pollen grains, cave stalagmites, and other newly discovered indicators of ancient climate, we now know that the glaciation was far from a monolithic event. Rather, Europe's climate shifted dramatically from one millennium to the next, in a constant seesaw of colder and warmer events that often brought near-modern climatic conditions to some areas. Old models assumed that Scandinavia was buried under huge ice sheets for all of the last glaciation. Now we know that this was the case only during the Last Glacial Maximum, about 21,500 to 18,000 years ago, when much of Europe was a polar desert. Much of the time Europe was far warmer, indeed near temperate. What is fascinating about the world of the Neanderthals and the Cro-Magnons is that we now have just enough climatological information to look behind the scenes, as it were, to examine the undercurrents of climate that caused hunting bands to advance and retreat and that perhaps helped drive some Neanderthal groups into extinction.

Cro-Magnon explores Ice Age societies both historically obscure and well known, not just within the narrow confines of Europe, but on a far wider canvas. The Cro-Magnons may have been Europeans, but they were comparative newcomers who arrived from elsewhere. We cannot understand them without journeying far from the familiar confines of Les Eyzies and the Cro-Magnon rock shelter. Thanks to mitochondrial DNA and Y chromosomes, we know that they were ultimately Africans. Rather startlingly, we also believe that humanity almost became extinct in the aftermath of a colossal explosion, when Mount Toba, on Sumatra, erupted into space about 73,500 years ago. Connecting the dots between dozens of archaeological sites is one of the exciting challenges facing the archaeologist of the future. Many of them are little more than scatters of stone artifacts, which we have to link to ash falls, to climate records wrested from cave stalagmites, to the fluctuations of the Sahara Desert, and to the harsh realities of a life lived in often arid or cold landscapes. All we have at the moment is a tentative framework, based on frequently inadequate data. But it is enough to allow us to peer at the late Ice Age world not from the outside, but from within, for the fundamental routines of hunting and foraging in arctic and tropical, semiarid environments remain much the same today as they were over twenty thousand years ago. There are only a few options for, say, hunting reindeer with spears, driving rabbits into nets, or trapping arctic foxes. We know of them from historic as well as still-living hunter-gatherer societies, whose basic subsistence activities have changed little over the millennia.

The story of the Neanderthals and the Cro-Magnons tells us much about how our forebears adapted to climatic crisis and sudden environmental change. Like us, they faced an uncertain future, and like us, they relied on uniquely human qualities of adaptiveness, ingenuity, and opportunism to carry them through an uncertain and challenging world. We have much to learn from the remote past described in these pages.

Author's Note

Geographical place names are spelled according to the most common usage. Archaeological sites are spelled as they appear most commonly in the sources I used to write this book. Some obscure locations are omitted from the maps for clarity; interested readers should consult the specialist literature.

The notes tend to emphasize sources with extensive bibliographies to allow you to enter the more specialized literature if you desire. This being a narrative account of the Cro-Magnons, sidebars in each chapter provide further information on technicalities such as radiocarbon dates, specialist controversies, and stone technologies.

All radiocarbon dates have been calibrated using the latest version of what is a constantly revised calibration curve. You can view the calibration curve at http://www.calpal.de.

A note on the use of the term *Cro-Magnon*: I use it in a generic sense in these pages, as it is a convenient, easily remembered term. Here it is employed interchangeably with *Homo sapiens*, *modern*, and *anatomically modern human (AMH)*. This is a literary compromise for clarity. Scientific reality is, of course, more complex and is fully explored in the specialist literature. Obviously, the Cro-Magnons themselves had no such equivalent term. From the beginning, they were a patchwork of bands, kin groups, and sometimes larger affiliations whose names have not come down to us. The point is, we gave these people their name, its origin a random moment in an extraordinary history.

Cultural terminology is always a thorny issue, especially with the Cro-Magnons, whose archaeology is dauntingly complex. Neanderthal societies flourished during the Middle Paleolithic and Cro-Magnon societies during the Upper Paleolithic, generic labels meaning Middle and Upper (or Late) Old Stone Age that I have not used in this book, although you will find them in the academic literature. Nor have I used the term *Mesolithic* (Middle Stone Age), which refers to hunter-gatherer societies after about ten thousand years ago. I have tried to keep often-arcane cultural terms to a minimum and, for the purposes of this narrative, have ignored the many subdivisions of the various cultures referred to in these pages. Most of them stem from stratigraphic observations of layers in archaeological sites and from differences in stone tools and other artifacts. While these are of vital importance to specialists, I do not consider most of them essential to this story.

CHAPTER 1

Momentous Encounters

THEY CALL HIM *LÖWENMENSCH*, "the Lion Man." The ivory figurine stands tall, leaning ever so slightly forward, arms by his sides. His head is a lion's, mouth slightly open, ears pricked, the mane cascading down the back. But his arms are human, relaxed, marked with six or seven striations (see color plate 1). The feet are slightly apart, a hint of maleness between the legs. The Lion Man stands serene, gazing calmly into the distance, contemplating an infinite landscape, a realm far beyond the confines of the living world. He came into being over thirty-four thousand years ago, carved out of water-soaked mammoth tusk by one of our remote ancestors, a Cro-Magnon.[1]

The artist who created the Lion Man was just like us. He laughed and cried, loved and hated, was calculating and sometimes devious. She was a member of a small hunting band, one of a few thousand people living in what is now southern Germany amidst a tapestry of coniferous forests and open tundra. Here, reindeer herds migrated north and south with the seasons. Great mammoths fed by icy streams; flocks of arctic ptarmigan croaked at water's edge. This was no Ice Age paradise. The Lion Man's creator lived in a world whose harsh realities included frequent hunger and savage winters. But it was also a realm of the mind's eye, peopled with vibrant animals and powerful supernatural forces, which formed symbolic partnerships between humans and beasts. *Löwenmensch*, with his leonine head and human limbs, bridged the chasm between the living and supernatural realms, the kingdom of the imagination. His maker drew on the awesome cognitive abilities we ourselves possess. Nimble and tall, the Cro-Magnons were identical anatomically and intellectually to modern humans. We know that their brains had an identical configuration to ours, that they were capable of articulate speech, just as we are.

The ancestors of the anonymous creator of the Lion Man had arrived in their challenging homeland about ten thousand years earlier from warmer and drier environments far to the southeast, in southwestern Asia. A new generation of radiocarbon dates tells us that the Cro-Magnons spread across Europe within a mere five thousand years. People moved constantly, responding to social needs and to intelligence about game, campsites, and water supplies. The distances across Europe from southwestern Asia seem enormous, but within a few generations, Cro-Magnon bands would have covered surprising expanses, especially in sparsely populated, often bitterly cold environments, where climatic conditions were constantly changing, often for a few years at a time, sometimes for several lifetimes, at other times seemingly permanently. It's easy to imagine population movements that spanned 250 miles (400 kilometers) within a generation or so. And wherever they settled, the Cro-Magnons encountered small bands of Neanderthals, the European indigenes, people with biological and cultural roots hundreds of thousands of years in the remote past.[2] About fifteen thousand years later, by about thirty thousand years ago, in one of the stunning developments of history, the Neanderthals were extinct.

These pages tell the story of the Cro-Magnons, beginning with their encounters with the primordial Neanderthals. The complex relationship between Cro-Magnon and Neanderthal has fascinated scholars for generations, as if it were the subject of an epic paleoanthropological novel. How did they perceive one another? Did they interbreed, or did the newcomers slaughter Neanderthals on sight? Were archaic and modern humans close neighbors, or did the Cro-Magnons simply push the indigenes out of their ancient hunting territories into marginal landscapes? Did the vastly superior intellectual abilities of the moderns play a central role in driving the Neanderthals into extinction, or were climate changes and extreme cold the ultimate villains? Reality, as far as we can know it, was far from an epic adventure. This is a story of brief but momentous encounters, of people separated by profound incomprehension and misunderstanding. It is also a tale not of great leaders or powerful warriors, but of ordinary Ice Age people rising to the challenge of surviving in brutal environments. What were the secrets of the Cro-Magnons' brilliant success? Was it their more-advanced technology, their hunting and foraging abilities, or brilliant innovation combined with opportunism? Or did their spiritual beliefs and complex relationship with the supernatural realm play a decisive role? The portrait of the Neanderthals and the Cro-Magnons and their world in these pages comes from cutting-edge multidisciplinary science and a growing knowledge of the dynamics of hunter-gatherer societies in every corner of the world. It's no exaggeration to say that the foundations of today's Europe were forged in the events of the late Ice Age, between about forty-five thousand and twelve thousand years ago.

WE MUST BEGIN by introducing the Cro-Magnons. In today's parlance, they are technically anatomically modern humans (AMHs). But the word *Cro-Magnon* rolls off the tongue much better and is a far more satisfying label for the first Europeans, even if it is technically somewhat incorrect. The name dates back to 1868, when the railroad came to the sleepy village of Les Eyzies, in southwestern France. Workmen clearing land for the new station uncovered a small, totally buried rock shelter and some flint tools and animal bones near a rock prophetically called Cro-Magnon, "great cavity." (I was disappointed on a recent visit to discover that there's nothing to see today except a small overhang behind a row of hotel workers' cottages and a weathered plaque on a rock wall.) A young geologist, Louis Lartet, dug into the back of the shelter soon after its discovery.[3] He unearthed five human skeletons, including the remains of a fetus and several adults, among them a woman who may have been killed by a blow to the head. The burials lay among a scatter of shell beads and ivory pendants. These were no Neanderthals with simple artifacts and no bodily decoration. The Cro-Magnon people had round heads and high foreheads and were identical to modern humans.

Les Eyzies lies on the bank of the Vézère River in a valley where high limestone cliffs with caves and deep overhangs provided wonderful shelter for Ice Age visitors. Louis Lartet's father, Édouard, had partnered with Henry Christy, a wealthy English banker, to dig into Les Eyzies' huge rock shelters in the early 1860s. They had uncovered flint artifacts, engraved harpoons, and numerous reindeer bones, but no human remains.[4] The Cro-Magnon find proved that the makers of these artifacts were *Homo sapiens*, the remote ancestors of modern Europeans, who lived during the Ice Age, during a period somewhat fancifully called l'Âge du Renne, or the Reindeer Age, because of the numerous bones of these animals found in the rock shelters. Soon scholars were comparing them (wrongly) to the Eskimo of the Arctic, but one fact was beyond question: they were the successors of the Neanderthals. Just where they came from is still the subject of lively academic debate.

The Cro-Magnons, among whom the creator of the Lion Man numbered, were but specks on a vast European landscape of deep river valleys, mountains, and boundless open plains. They were well aware they were not the only humans preying on bison and reindeer, seizing meat from predator kills, stalking wild oxen on the edges of dark green pine forests. Just occasionally, they would glimpse their rivals—a Neanderthal band slipping quietly across a water meadow, people so different that Cro-Magnon children would run away. Like the Neanderthals, the Cro-Magnon newcomers were thin on the ground. But they were completely different. They were *Homo sapiens*, "the wise person," capable of flexible thinking, planning ahead, and fully articulate speech. Europe was never the same after their arrival.

On long winter nights, the older men, perhaps those with unusual supernatural powers, would tell stories of a time long ago when their exotic neighbors were thicker on the ground. But even then they were a rare presence, glimpsed walking quietly among the trees or high above a valley on a steep hillside. Now there were far fewer of them. Close encounters were an unusual event, perhaps during a hunt, or when collecting honey in the summer. Perhaps two handfuls of men and boys out hunting would face off unexpectedly, spears in hand, watching closely for a threatening gesture. The physical contrast was dramatic: tall, slender Cro-Magnons; compact Neanderthals.[4] The Cro-Magnons wore close-fitting fur parkas, long pants, and waterproof boots. Their potential adversaries were barefoot men of immense strength, their bodies draped in thick furs crudely joined with thongs. They carried heavy, fire-hardened spears and wooden clubs, nothing more, weaponry virtually identical to that carried by their remote ancestors tens of thousands of years before. Each side would stare at the other. Perhaps a few gestures would ensue, universal to all humans: a smile, a proffered gift of a honeycomb, perhaps some quiet grunts. There was no shared language, perhaps not even a common body odor. After a few moments, Cro-Magnon and Neanderthal would likely go their separate ways. We can only guess at the nature of such encounters. Our only potential analogies come from meetings between Western explorers and hitherto unknown societies, like, for example, the Tasmanian Aborigines, in the late eighteenth century. The Tasmanians had encountered no outsiders since rising sea levels had isolated them from mainland Australia nine thousand years earlier. Both sides recognized the other as fellow humans, but beyond that and some common gestures of friendship like a smile, they lived in entirely different worlds. We can be sure that any brief meetings with Neanderthals would be long remembered by the moderns, who would pass on recollections of such unusual events from one generation to the next. One is reminded of the New Zealand Maori, who still retained vivid memories of Captain James Cook and his ships a century after he departed over the horizon.

CENTRAL EUROPE, FORTY thousand years ago. Thick mist lies low over the stream, obscuring the tops of the dark fir trees. The fast-flowing water riffles loudly in the still air. A reindeer with magnificent antlers grazes quietly by the riverbank, knee-deep in the cold shallows. Two Neanderthal hunters watch silently, crouched in the snow, hidden by large boulders. They have been stalking the reindeer since first light, oblivious to the biting cold, slipping quietly from tree to tree in the dark shadows. Now they are in range, only a few feet from their prey. The older man slowly raises his fire-hardened wooden spear and crouches for a quick leap for the reindeer's vulnerable heart. His prey looks up unexpectedly, alert to a virtually inaudible sound on the far bank. *Phut! Phut!* Moments later, two antler-tipped spears cast from among the trees land on target. The beast staggers and falls into the river. An eddy carries the still-twitching carcass across a deep pool to the other bank.

As the dead reindeer floats out of range, the Neanderthals lower their weapons. Just for a moment, the mist parts. Three men in fur parkas look down at the fitfully moving beast, long spears and spear throwers in hand. As the two younger hunters drag the reindeer from the water, the oldest man shakes loose the hood of his parka, revealing a head of thick red hair above a smooth, well-rounded forehead. He glances dismissively at the skin-clad Neanderthals and raises his spear in contemptuous defiance. Moments later, the strangers vanish into the forest with their prey . . .

The Neanderthals, *Homo neanderthalis*, require no introduction to any reader of popular science. They are the cave people of prehistory, the hirsute folk with wooden clubs, a grossly unfair characterization of skilled, tough hunters armed with little more than wooden thrusting spears, who were not afraid to hunt such formidable beasts as the European bison and the aurochs, *Bos primigenius*, the fierce primordial wild ox. But there is much more to the Neanderthals than cartoonists' stereotypes. Forty-five thousand years ago, perhaps fifteen thousand to twenty thousand of them lived between the Atlantic Ocean in the west and the Ural Mountains, in Eurasia, far to the east. They hunted and foraged in small family bands. Most of them encountered no more than a few dozen fellow humans during their lifetime and then only briefly, perhaps for a cooperative hunt or to obtain a mate. They thrived for thousands of years in some of the most brutal environments of the late Ice Age world, including extremes of cold—when they had Europe to themselves (figure 1.1).

Where did they come from? What we know of their early history has been gleaned from meager scatters of stone tools and a few human fossils. These tell us that Neanderthal ancestry goes back far into remote prehistory, described in chapter 2. The earliest definite Neanderthals date to about 200,000 years ago. We also know, from an increasing number of archaeological sites, that their numbers rose slowly after about

Dates are in approximate calendar years ago, mostly calibrated radiocarbon readings. Mya = million years ago.

10,000	Increasingly diverse hunter-gatherer societies throughout Europe.
11,000	Magdalenian culture ends.
c.12,000	Agriculture and animal domestication in the Near East.
13,000	Niaux paintings.
14,800	Altamira paintings and engravings.
17,000	Magdalenian groups expand northward as western Europe warms.
17,000(?)	Lascaux paintings and engravings.
18,000	Magdalenian culture begins.
	Solutrean groups in northern Spain and southwestern France.
21,500/ 18,000	Last Glacial Maximum.
25,000	Heyday of the Gravettian culture, which persists in various forms in much of eastern Europe until the end of the last glaciation.
30,000(?)	Neanderthals become extinct.
29,000	Aurignacian culture ends.
32,000?	Grotte de Chauvet paintings—date range uncertain.
39,000	Aurignacian culture appears over much of Europe.
	Campanian eruption in Italy.
42,500(?)	Cro-Magnons appear in western Europe.
45,000	Modern humans at Kostenki in eastern Europe, having spread north from the Near East.
c.55,000	Fully modern humans spread out of Africa. Exact date is uncertain.
70,000	End of megadroughts in Africa.
	First Last Glacial Maximum in Europe.
73,500	Mount Toba eruption decimates humanity.
	Modern humans become extinct in the Near East (?).
100,000(?)	Some *Homo sapiens* groups settle in the Near East from Africa.
128,000/ 115,000	Last Interglacial.
160,000	Herto *Homo sapiens* fossils, Ethiopia.
171,500	Genetic estimate of the appearance of the first modern humans.
195,000	Omo Kibish *Homo sapiens* fossil, Ethiopia.
200,000	Neanderthals well established in Europe.
400,000	Atapuerca *Homo heidelbergensis*. Possibly ancestral Neanderthals.
500,000	Mauer *Homo heidelbergensis,* Germany.
600,000	*Homo heidelbergensis* evolves in Africa (?).
1.2 mya	Human settlement at Sima del Elefante, Atapuerca, Spain.
1.6 mya	Dmanisi fossils, Georgia.
1.8 mya	*Homo ergaster* moves out of Africa (?).
2 mya	*Homo ergaster* evolves out of earlier *Homo*.

Figure 1.1 *Major developments and events covered in this book.*

150,000 years ago, following a period of intense cold that lasted at least 30,000 years. They flourished, albeit in smallish numbers, through subsequent, more temperate millennia. In what is now Italy, they hunted elephant and hippopotamus about 125,000 years ago. The warmer conditions lasted until about 115,000 years before the present, when the last glacial period of the Ice Age brought much colder temperatures and major environmental changes. By then, Neanderthals thrived in small numbers over an enormous area of Europe and Asia, from southern Britain and the Atlantic, through Belgium and France, across central Europe, and deep into central Asia, far east of the Black Sea into modern-day Uzbekistan and perhaps beyond. Neanderthal bands flourished in warmer environments, too, in the Near East in Greece, and in Spain as far south as Gibraltar (see figure 3.2).

For all their wide distribution, the small populations of the Neanderthals were a fleeting presence in grand Ice Age landscapes. They were the only human beings in a dangerous, predator-rich world, where survival depended on careful observation, constant watchfulness, and opportunism. Theirs was a rhythm of life that shifted infinitesimally over tens of thousands of years until the Cro-Magnons invaded ancient Neanderthal hunting territories and disturbed the even tenor of their days.

SUCH, THEN, ARE the protagonists of our story, the indigenes and the newcomers. Édouard Lartet and Henry Christy established that the Cro-Magnons were the successors of the Neanderthals. But where had the Cro-Magnons come from? Had they originated in Europe itself, as many early Eurocentric scholars assumed? Or had they come from elsewhere? Experts pointed to the densely packed layers of rock shelters and caves in the Vézère Valley and northern Spain, to the seemingly orderly transition of artifacts from one stratified layer to another over thousands of years, starting with Neanderthal occupation and ending up at the end of the Ice Age. However, as archaeological research expanded into regions like the Danube Valley and the Near East before World War II, it became apparent from different artifact sequences that the Cro-Magnons were outsiders who had originated far from western Europe.

If ever there was an archaeological will-o'-the-wisp, the search for the first modern humans is one. The quest has shifted from the relatively familiar caves and rock shelters of the Near East, until recently the favorite candidate for place of origin, into the depths of tropical Africa. I worked in southern Africa early in my career, in the 1960s, and had a chance to examine many archaeological collections in the museum in Cape Town and elsewhere. The artifacts were nothing spectacular, mostly trimmed flakes struck off from stone nodules, scraping tools, occasional stone projectile points, and, from later caves and shell heaps, enormous quantities of small stone arrow barbs and numerous bone tools. There were few radiocarbon dates; most of the excavations dated to the 1930s. When I was there, only a handful of excavations existed between Sudan and the Cape of Good Hope, but the numbers have increased dramatically since the 1970s. New digs have transformed our knowledge of Africa over one hundred thousand years ago. We now know that *Homo sapiens* flourished in tropical Africa long before modern people colonized Europe or the Near East.

The first strong clues came during the 1980s, when some fragments of modern humans came from a one-hundred-thousand- to eighty-thousand-year-old occupation level in one of the Klasies River caves, on the southeast African coast. These were the earliest *Homo sapiens* finds in the world at the time, and I, among others, had trouble accepting the chronology. At this point, molecular biologists studying mitochondrial DNA (mtDNA), which is inherited through the female line, threw an intellectual cat among the proverbial pigeons. A group of geneticists headed by Rebecca Cann and Alan Wilson, using mtDNA and a sophisticated "molecular clock," traced modern-human ancestry back to isolated African populations dating to between two hundred thousand and one hundred thousand years ago. Inevitably there was talk of an "African Eve," a first modern woman, the hypothetical ancestor of all modern humankind. Most archaeologists gulped and took a deep breath. Cann and her colleagues had taken *Homo sapiens* into new and uncharted historical territory.[5]

Furious controversy surrounded the African Eve, pitting biological anthropologists who believed that all modern humanity had originated in Africa against those who argued for multiple origins in different parts of the Old World. As we will see in chapter 5, molecular biology is now much more refined, and the mtDNA, and now Y chromosome, samples are larger. The genetic case for an African origin for *Homo sapiens* seems overwhelming. The archaeologists have also stepped forward with new fossil discoveries, including a robust 195,000-year-old modern human from Omo Kibish, in Ethiopia, and three 160,000-year-old *Homo sapiens* skulls from Herto, also in Ethiopia. Few anthropologists now doubt that Africa was the cradle of *Homo sapiens* and home to the remotest ancestors of the first modern Europeans—the Cro-Magnons. The seemingly outrageous chronology of two decades ago is now accepted as historical reality.

If *Homo sapiens* indeed originated in tropical Africa, how and when did the descendants of whom we can, somewhat indulgently, call the African Eve move into the semiarid lands of the Near East? Here we embark into the realm of speculation, largely because small bands of hunter-gatherers leave few traces of their passing behind them. We're back with the archaeological will-o'-the-wisp, forced to rely on general clues. We can be pretty certain the migrants were people adapted to open country, who were constantly on the move. Their tools and weaponry had to be carried everywhere, so it is hardly surprising that little survives except for occasional groupings of stone tools.

A generation ago, we thought of a single, albeit complex, movement out of Africa, perhaps about one hundred thousand years ago. This relatively simple model has given way to a more complex scenario involving two out-migrations. The first may indeed have occurred about one hundred thousand years ago but seems to have fizzled in the Near East, perhaps in the face of drought. A second, even less well-documented push seems to have taken place later, around fifty thousand years ago. This time, moderns settled throughout Near East Asia and stayed there, apparently living alongside a sparse Neanderthal population. This widely accepted theory assumes that by this time the newcomers had all the intellectual capabilities of *Homo sapiens*. Just when and how they acquired them remains a major unsolved problem. All we can say is that at some point between one hundred thousand and fifty thousand years ago, at a seminal yet still little known moment in history, *Homo sapiens* developed the full battery of cognitive skills that we ourselves possess. After a surprisingly short time, perhaps a mere five thousand years, their descendants moved northward into Eurasia and Europe.

Like most archaeologists, I have a profound distrust of theoretical scenarios without the sites and artifacts to back them up, but in the case of modern-human origins, we have to work with such tools in the absence of much hard data. As we shall see in chapter 5, the two out-migrations theory seems the most convincing working model for bringing the ancestors of the Cro-Magnons into their new homeland. Refinements, indeed wholesale changes, are likely to descend on this model in future years, but one fundamental point is of great importance: in the final analysis, the first modern settlers of Eurasia and Europe were ultimately Africans. Many of their hunting practices, light weapons, and social institutions developed in semiarid lands south of the Sahara Desert in the tropics. I believe this ancestry had a profound influence on the ways they adapted their lives to a much colder world of climatic extremes.

THE ENCOUNTERS BETWEEN Neanderthals and Cro-Magnons in Europe and Eurasia were an intricate social gavotte that played out over many centuries. There was nothing new about the dance, for the newcomers had met Neanderthals many times before. Their ancestors had lived alongside Neanderthals in the semiarid lands of southwest Asia. You can be sure that oral traditions of their dealings with what to them must have been somewhat alien beings passed down through the generations. The Neanderthals were exotic because they lacked a common tongue, if, indeed, they possessed fluent speech at all. By the time the first Cro-Magnons arrived in Europe, a huge intellectual and social chasm separated them from their neighbors. We will, of course, never know what they thought of the Neanderthals. They may have respected their great strength and their stalking abilities, but I suspect the Cro-Magnons thought of the Neanderthals not as humans like themselves, but as something that resembled them but acted, smelled, and spoke entirely differently. Each may well have avoided the other, for they had nothing in common. Their encounters were likely mostly momentary contacts in sheltered river valleys and on open plains, by lakes and on seacoasts, by rock shelters and in caves. Archaic and modern lived alongside one another, probably at a distance, until the last Neanderthals died out, probably in Spain, some thirty thousand years ago; the date is controversial.

A huge academic literature surrounds the Neanderthals and their fateful encounters with the aggressive and opportunistic newcomers. Theories abound, as do questions, most of them virtually impossible to answer. Did modern humans attack their new neighbors and rapidly drive them out of their favorite hunting territories into marginal environments? To prove this would require dozens of regional maps with site distributions of numerous accurately dated Neanderthal and Cro-Magnon sites. Unfortunately, we don't have the sites, let alone a way of dating them accurately to within the span of a few generations, which is the kind of precision one would need. Take another scenario. Did Cro-Magnon bands kill off Neanderthals whenever they encountered them? Once again, the proof is near impossible to acquire. You would need to find human skeletons with spear points embedded in them—not just a single burial but dozens of them in different locations. So far, we have none. Then there's sex, a thorny subject that provokes news headlines without fail. Did Neanderthals and Cro-Magnons interbreed? Some years ago, the geneticists Svante Pääbo and Matthias Krings of the University of Munich succeeded in extracting a partial DNA sequence from a Neanderthal limb bone. Recently, Pääbo and his colleagues decoded most of the Neanderthal's mtDNA.[6] They found that the Neanderthal sequence falls outside the range of genetic variation in modern humans, which means they were not direct ancestors of *Homo sapiens*. Humans and chimpanzees share over 98 percent of their DNA sequence. Neanderthals were even closer to moderns, but the small differences are enough to show that we began to diverge from them around seven hundred thousand years ago. Could Cro-Magnons and Neanderthals then have interbred? Most experts think they did not.

There remains the most popular theory: modern humans were simply more adept at hunting and survival in very challenging, ever-changing late Ice Age environments. You can point to the Cro-Magnons' superior weapons and more efficient technology, to their clothing, and, above all, to their enhanced cognitive abilities. All of these must have been players in the ongoing gavotte, but to invoke a single, overriding cause for the Neanderthals' demise is to court accusations of oversimplification. Here, however, we can make a stronger case, based on both archaeological finds and intelligent speculation.

Both Neanderthals and Cro-Magnons coped effortlessly with abrupt climatic changes from near-temperate to extremely frigid conditions. How well, however, the Neanderthals were able to deal with deep snow cover and long months of subzero temperatures is a matter of ongoing debate. They lacked what was, perhaps, one of the most revolutionary inventions in history, and an inconspicuous one at that: the eyed needle, fashioned from a sliver of antler, bone, or ivory. If their expertise with antler is any guide, the Cro-Magnons must have been adept woodworkers in the more temperate environments of southwestern Asia. When they moved north, they settled on a continent where antler and bone were potential replacements for wood, and where mammoth and other large animal bones had to be used as fuel in more treeless environments. With brilliant opportunism, they used small stone chisels to remove fine splinters from antler and bone, which they then ground and polished into slender needles. Carefully fashioned stone awls served as drills to make the holes for the thongs that served as thread, substitutes for the vegetable fibers used with wooden needles in their ancestral homes.

Every Cro-Magnon, man, woman, and child, must have been aware that protection from clothing came in layers, that warmth escaped from the head and extremities. As we will see, an indirect source of information on the garments they wore is the traditional clothing used by Eskimo and Inuit in very cold environments—the argument being that there are only a limited number of ways in which layered, cold-weather clothing can be fashioned from hides and skins. The needle allowed women to tailor garments from the fur and skin of different animals, such as wolves, reindeer, and arctic foxes, taking full advantage of each hide or pelt's unique qualities to reduce the dangers of frostbite and hypothermia in environments of rapidly changing extremes.[7] We cannot overestimate the importance of tailored clothing in Cro-Magnon life, especially when stacked up against draped skins.

Cro-Magnon hunters also relied heavily on more lethal, lighter-weight stone-tipped spears with greater range, more effective weaponry than the fire-hardened weapons of their neighbors. In the long run, two innovations, layered clothing and more effective projectiles, gave the Cro-Magnons a decisive practical advantage over the Neanderthals. The one enabled them to hunt efficiently in extremely cold, changeable conditions. The other allowed them to harvest a wider range of animals, especially medium- and smaller-sized beasts, which provided not only meat but also furs and other vital commodities for survival. They also used ingenious devices for procuring small mammals, birds (including waterfowl in the late spring and summer), and eventually fish: nets, traps, light throwing darts, and so forth. Cro-Magnon technological imagination opened up for them a whole new ecological niche that had been beyond the reach of their predecessors.

Was there another decisive advantage? The answer is probably an emphatic yes. What gave the newcomers the real edge was their intellectual awareness and imagination, their ability not only to cooperate with others, as the Neanderthals did, but also to plan ahead and to think of their surroundings as a living, vibrant world. This they defined with art and ritual, ceremony, chant, and dance, which helped them ride out the punches of rapid climate change and brutal temperatures, occasional hunger, and catastrophic hunting accidents. Their imaginations, their rituals, gave them a far more important cushion against harsh environments than any technological devices. We know from their art that they looked at their world with more than practical eyes, through a lens of the intangible that changed constantly over the generations. It was this symbolism, these beliefs, as much as their technological innovations and layered clothing, that gave them the decisive advantage over their neighbors in the seesawlike climatic world of the late Ice Age. There were more of them living in larger groups than there were Neanderthals, too, so there were more intense social interactions, much greater food gathering activity from an early age, and an ongoing culture of innovation that came from a growing sophistication of language, advances in technology, and a greater life expectancy. In a world where all knowledge passed orally from one generation to the next, this enhanced cultural buffer between the moderns and the harsh climate provided an extra, albeit sometimes fragile, layer of protection during the intense cold of the so-called Last Glacial Maximum, from 21,500 to 18,000 years ago.

There remain many uncertainties and profound disagreements over these various theories, but what *do* we know? There are no certainties, just some possible realities, which are a matter of instinct and extrapolation rather than scientific fact. I think we can safely assume that most contacts between the two groups were sporadic and of short duration. Each side would have known the other was there—watching, competing for game, and wary of attack, sudden contact, or simply loud shouting. The Cro-Magnons may have considered the Neanderthals, with their heavy brows and massive shoulders, repulsive in appearance—and vice versa. The contacts must have been like shadowboxing, with occasional sightings, sometimes little more than a presence glimpsed briefly among the trees.

I think we can assume that the two groups were intensely curious about each other, both as potential competitors and as fellow humans, however different, with their own arcane knowledge and unique skills. We can be certain, too, that an intellectual void separated Neanderthal and newcomer. The Neanderthals still lived in the same simple way as their remote ancestors, in an annual round that changed infinitesimally over hundreds of generations. They lacked fully articulate speech and had few of the intellectual abilities of the newcomers. We do not even know if they believed in an afterlife, although they did sometimes bury their dead. They communicated with gestures and sound in ways honed over hundreds of thousands of years, but with nothing like the sophistication and panache of *Homo sapiens*. The Neanderthals were indeed human beings, but they lacked the humanity of the Cro-Magnons. Unlike the Neanderthals, the Cro-Magnons celebrated, and sought to understand, their world with the help of their thoughts and imaginations, in chant, oral tradition, and ritual.

Throughout Europe and Eurasia, the two peoples lived alongside one another for many generations, not necessarily competitors or afraid of one another, but always careful to observe each other's doings. The Neanderthals would have been the more silent presence, usually on the margins, usually invisible. They would never have been out of the Cro-Magnon mind.

We can imagine a bright Cro-Magnon hearth in the heart of an open campsite on the plains in high summer, skin tents pitched on the perimeter. The people are dancing to a drum and flute, their profiles shimmering in the long shadows of the hearth. They are under the spell of the dance, oblivious to the small group of Neanderthals, who are watching silently and invisibly just outside the circle of tents and firelight. When the dance ends, they will slip away without a sound, yet subconsciously the Cro-Magnons know they are there. There must have been a form of silent modus vivendi between modern and premodern, a tolerance based on incomprehension, yet a realization that the one had something to offer the other . . .

WHAT WAS IT like to live alongside a totally different human species? We will never know, because *Homo sapiens* now represents all humankind on earth. Our only experience comes from the age of Western discovery, from colonial expansion and conquest, which is hardly a reliable comparison, except in one sense: that of a mutual lack of understanding. There've been a few times in my life when I've encountered people who lived in a completely different way from me: a San hunter from the Kalahari Desert, a village of subsistence farmers in Malawi. We had nothing in common, not artifacts, diet, dwellings, or a language. I remember gazing at them silently, conscious that we lived in completely different worlds. All we had in common were some universal gestures: a smile, a frown, and a pointed finger. It was both disconcerting and frightening.

As we will see in chapter 5, the San hunter-gatherers of southern Africa, famous for their magnificent cave art, have an ancestry deep in the past, so much so that we can compare their situation to that of the Neanderthals. They are, of course, modern humans, so the analogy is somewhat far-fetched, except in examining the situation they faced about two thousand years ago. About the time of Christ, Bantu-speaking farmers and cattle people spread rapidly from East Africa into the ancient San homelands. The two lifestyles—agriculturalist and hunter-gatherer—were totally incompatible, just like those of the Neanderthals and Cro-Magnons. Inevitably, the San suffered the most. Some of them may have adopted the new economies and become farmers and cattle people. Others may have perished. Many groups moved out to the margins of settled farmlands and continued to live as they always had. And there they remained for many centuries.

The farmers were well aware of the San. We know from oral traditions that many San communities exchanged honey, animal skins, and game meat for grain and other commodities in relationships that endured for generations. The Lala farmers of what is now eastern Zambia traded regularly with the San, whom they called *utunuta mafumo*, "the people with wrinkled skin on their stomachs that covers their genitals."[8] (For some reason, this label causes many African audiences to dissolve into hysterical laughter.) The *utunuta mafumo* looked different. They were shorter of stature, hard to see, and vain—people outside the normal parameters of Lala life. The Lala's interactions with them reflected an acceptance that there were others in their world, albeit at its margins—and this may have been the way that the Cro-Magnons viewed their neighbors.

The Lala and the *utunuta mafumo* distrusted one another. The two groups had no language in common, nor did they obtain food in the same way. Neanderthals and Cro-Magnons would have faced even greater hurdles of comprehension. They might have had some basic hunting skills and environmental knowledge in common, but the two groups could only communicate by gestures and signs, signs of friendship or enmity, or by offering gifts in cautious exchange.

What about communicating with people who have no experience of strangers? Few people living today have ever faced the problem of interacting with people who have absolutely no knowledge of the wider world. During the late 1960s, the famed BBC traveler David Attenborough visited the headwaters of the Sepik River, in New Guinea, where there were reports by passing aircraft of dwellings in rugged country that was thought to be uninhabited.[9] After two weeks of marching through remote terrain, Attenborough's small party came across two sets of footprints. They followed them, laying out gifts in the forest and offering greetings in known river dialects. The strangers left telltale spoors that showed they were watching constantly. The Europeans lost the trail and were about to give up their search when suddenly seven small, almost-naked men appeared in the bush near their camp. Attenborough recounts how he and his companions made hasty gestures of friendship, for the tribesmen did not understand any known river dialect. Fortunately, the two sides had gestures in common: smiles and eyebrows that could frown, signify wonder or disapproval, or ask a question. The gestures were the only way to understand one another, to barter iron knives for food, and to deepen the relationship between two absolutely incompatible societies.

How DID THE Neanderthals and Cro-Magnons make contact other than by accidental encounter? They lived on the same continent, but they inhabited very different worlds, the one, as had always been the case, defined by the fundamental rhythms of the seasons, the other by the same endless cycle, but by much more as well. Their worlds and daily routines may have varied dramatically, but no hunting society is ever completely self-sufficient. The Neanderthals had an ancient, and profound, knowledge of what at first was unknown territory to the newcomers. They would have known where to find such esoterica as honeycombs, maybe supplied bearskins or hides to their neighbors, receiving perforated bear teeth or some tools and game meat in exchange. Ideas and hunting intelligence may have percolated between Neanderthal and newcomer as they watched each other. In chapter 6, I describe how some Neanderthal bands in France attempted to copy artifacts made by their more sophisticated neighbors. Given that the only common mode of communication between the two may have been gestures or grunts, perhaps some form of "silent trade" may have brought them together.

Silent trade was relatively commonplace throughout the world as recently as the last century. The Greek historian Herodotus described such transactions between the Scythians of Eurasia and their neighbors. In 1511, a Portuguese convict, Antonio Fernandez, was sent from the Indian Ocean coast to explore the interior of southeast Africa. He reported on the copper trade between Arab merchants and local people along the Zambezi River: "These men are badly proportioned, and are not very black, and they have tails like sheep." The locals would carry ingots across the river, lay them out on the bank, and then withdraw. The traders would inspect the shipment. If they were satisfied, they would "leave the cloths and any other merchandise that they carry."[10] The back-and-forth would continue until both parties were satisfied, yet not a word was exchanged.

We can imagine another stream, another place, bounded by dense conifer forest, an unspoken boundary between a Cro-Magnon band and some Neanderthals. Two or three squat figures appear, carrying honey in skin containers and some bearskins. They lay them on a boulder on the far side of the stream, then withdraw into the trees and wait. Some time later, a few watchful Cro-Magnons ford the river, spears in hand. They examine the skins and taste the honey, licking their hands. One of the men lays out some perforated teeth and a string of seashells. Then they retreat, vanish from sight, and watch. The slow-moving exchange takes hours, even days, as the unspoken bargaining continues, hide by hide, artifact by artifact. Eventually each side is satisfied and gathers up its share, perhaps even looking across at the other silently before vanishing into the forest shadows . . .

For all their differences, Neanderthal and Cro-Magnon had to confront that most perennial of issues for people with large hunting territories who were on the move for much of the year: how do you deal with strangers? The answer must have been—very cautiously.

BY THE STANDARDS of later times, the gavotte between Neanderthals and Cro-Magnons was history in slow motion, involving tiny numbers of people on both sides. Above all, it was a series of accidental encounters. Like so many other major events in human history, the arrival of the Cro-Magnons in Europe and Eurasia was not a deliberate act, planned for months and executed with precision, as if an army of hunters were on the march. Rather, modern humans arrived in dribs and drabs in Eurasia and what is now the Ukraine, on the banks of the Danube River, and eventually in the sheltered valleys of southwestern France. They settled along Mediterranean shores and as far west as northern Spain. Their movements consumed many, albeit short, generations, part of the endless population shifts of hunter-gatherer groups exploiting large hunting territories where food supplies were patchy and often widely separated. The number of people involved was infinitesimal, but their biological and cultural advantages over the indigenous population were enormous.

Thanks to remarkably precise radiocarbon dates, we know that the Cro-Magnon move into the heart of the Neanderthal world came about haphazardly about forty-five thousand years ago. Small family bands followed migrating game into seemingly uninhabited lands. They found sheltered places to winter and locations where nuts were plentiful during the brief autumn. All kinds of trivial events triggered the expansion, all of them part and parcel of daily life: Sons split off from their families; bands fractured and dispersed when men or women quarreled with their siblings; people married into neighboring bands; hunters died in violent accidents, and their families joined other groups. The ebb and flow of people, of groups, fellow kin, families, and individuals, never ceased, propelled by the realities of the foraging life and its profound flexibility. And, time and time again, Cro-Magnon encountered Neanderthal, with inconspicuous but ultimately momentous consequences for history.

As we shall see, the initial settlement of Europe and Eurasia took place at an opportune time, when the climate to the north and west of the eastern Mediterranean was briefly warmer than it had been for a long time. The process of colonization was seemingly rapid, perhaps occupying but some five thousand years. But the Cro-Magnons' initial steps into a harsh and unknown world disrupted and eventually destroyed a form of human existence that had remained virtually unchanged for more than two hundred thousand years, its ancestry even deeper in the remote past, where our story begins. In its place, the newcomers created entirely new Ice Age societies, in which we can discern a distant mirror of ourselves.

CHAPTER 2

Neanderthal Ancestors

WHERE DID THE NEANDERTHALS ultimately come from? The search begins not in Europe but in the vastness of the African savanna, where the first humans flourished 2.5 million years ago.

Parts of Africa are little changed since those remote days. You never forget Tanzania's Serengeti Plain, where the horizon stretches to infinity and the pale blue sky arches high overhead. Here you witness the African savanna much as it was two million years ago. The Serengeti is an endless, gently rolling plain of short, brown grass, interspersed with umbrella-like acacia trees, whose thin leaves provide a tracery of inadequate shade. Shallow gullies dissect the grassland with muddy water holes where wildlife congregates at dusk. Harsh, unforgiving, and arid for months on end, this seemingly inhospitable world supports a rich and varied bestiary that has changed little since the time of the first humans. Here elephants wallow at water's edge and giraffes browse on thorny trees. Tens of thousands of wildebeests migrate across the plain, like a forest on the move. And if you look closely, you'll almost certainly see a pride of lions resting in the shade or lacerating the carcass of a recent antelope kill, while hyenas await their turn and vultures hover and wheel overhead.

As the late Harvard biologist Stephen Jay Gould once memorably remarked, we humans are all descendants from the same African twig. And it was here, on the African savanna, that the story of Europe, and of the Neanderthals, ultimately began. Here the remote ancestors of the first European Neanderthals learned how to hunt big game, virtually their only weapons a long wooden spear and an opportunistic ability to cooperate with others. They lived in a state of constant watchfulness, alert for the predators that lurked on every side, relying on nimbleness and ingenuity for protection, knowing that their animal enemies were wary of bristling weapons. Their success at the hunt changed history. Less than a million years later, a few handfuls of premodern people crossed the Sahara and western Asia and settled in Europe, taking with them hunting skills honed on the semiarid savanna.

Why should we look to Africa? The easy answer is that this is where humanity evolved, but that does not tell us why small numbers of premodern people crossed the Sahara and entered the Mediterranean and European worlds. Climate changes may have played a role, for the story of premodern Europe began about two million years ago, when Africa became progressively drier. The climatic shift had actually begun some five million years earlier, as tropical forests slowly gave way to more open scrub savannas, where seasonal aridity favored grasses and shrubs and a loose scatter of trees. These extensive, more open, and seemingly unproductive landscapes in fact provided accessible, more nutritious, and palatable food for herbivores of all kinds. People pursued these beasts. And these herbivores in turn provided food for humans.

As the savanna spread in the face of more arid conditions, so did populations of herbivores, among them elephants, antelope, zebras, and rhinoceroses. These mammals became the dominant large animals of Ice Age Africa, Europe, and other parts of the world. They thrived in dry, seasonal environments because they consumed massive amounts of vegetable tissue, which required complex digestive tracts and large bodies. Antelope and other herbivores also ingest fiber-digesting microorganisms that process fibrous tissue in their guts. The large bodies that many of them have reflect their very active digestive systems. As the paleontologist R. Dale Guthrie remarks of the African eland, a large antelope, "the complex digestive physiology required to convert a coarse bit of gray shrub into a healthy eland is a little like trying to make cheese from chalk."[1]

The humans of two million years ago were not herbivores. People lived, for the most part, in open country and solved the problem of survival on the savanna by consuming the herbivores, which had digested the flora. They collected many plant foods, not only grasses and fruit but also tubers. However, meat was the most important staple in the harsh, seasonal environments of the more arid lands. While it may have been the ideal food, collecting it by scavenging from predator kills was fraught with danger. The story of Europe begins in Africa not only because the remote ancestors of the Neanderthals evolved there, but also because that was where people learned how to hunt large animals.

WHO WAS THIS first hunter? The trend toward greater aridity coincided with the appearance of a new species of human, known to scientists as *Homo ergaster* (figure 2.1). Very different from and much more archaic human ancestors than the much later Neanderthals, these people stood upright, were fleet of foot, and had basically modern limbs, an elongated head with a strong browridge, and a brain capacity about three quarters of that of modern humans.[2] *Homo ergaster*'s anatomy reflected a life that involved covering long distances in open country. The species was larger than their predecessors, *Homo habilis*, too, the males weighing as much as 130 pounds (60 kilograms). The newcomers matured more slowly than apes and enjoyed a longer childhood, as well as a greater life expectancy, perhaps in their early twenties, although a much shorter one than that of modern people. Most likely, they lived in relatively small social groups, which ranged over large distances. Highly mobile and very watchful, they constantly acquired information about food and water supplies over many square miles of open country.

Homo ergaster soon colonized large areas of Africa. Like the herbivores that were their prey, the hunting bands migrated with changing vegetation zones and moved alongside predators like lions, leopards, and hyenas, animals with whom they shared the ability of opportunistic hunting.

The Neanderthals were expert hunters, but when did hunting begin? Once again, the answer lies in Africa. *Homo ergaster* was an omnivore, completely accustomed to quite drastic environmental changes in the distribution of open grassland, forest, and semiarid terrain and the dietary shifts that went with them. Unlike their predecessors, these people were serious hunters and meat eaters—because they dwelled for the

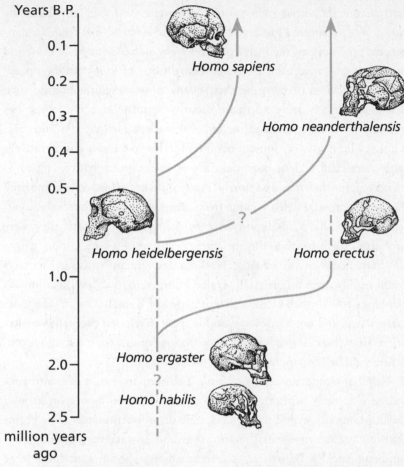

Figure 2.1 *A greatly simplified diagram of human evolution after two million years ago, showing the putative relationships between* Homo ergaster *and later human forms. The ancestry of both the Neanderthals and* Homo sapiens *lie with much earlier peoples.*

most part in open country, where meat was the dominant, though not, of course, the only, food source. We know this because the bones of numerous large mammals appear alongside stone butchering tools in some of the archaeological sites that document their wanderings, whereas none appear in sites that predate them.

Living on the savanna was always hazardous, especially for relatively frail mammals whose only protections were their ingenuity and the simplest of weapons. Hunting even a medium-sized animal could be dangerous; predators awaited easy prey on every side. Antelope, zebras, and other animals protect themselves by being fleeter of foot than their predators, by kicking, or by using their horns. *Homo ergaster* had no such defenses, nor any truly adequate security against large predators. No amount of shouting, stone throwing, or threatening behavior would deter a hungry lion, so every human band had to live extremely conservatively, with a constant eye for potential danger. Every group had young to be protected, for they matured slowly. They could not afford to become prey. The only potential defense came from weapons, objects that were, in effect, akin to sharp antelope horns. In their simplest forms, they were probably little more than thorny acacia branches or a long sapling with a tip sharpened with a stone flake. Half a dozen humans with such artifacts could readily deter a lion. As R. Dale Guthrie remarks, "even Hannibal's elephants bolted when faced with lethal rows of Roman pikes."[3] The long, razor-sharp, and fire-hardened wooden spear developed over many generations from these simple beginnings to become one of the most significant, if rarely studied, weapons in human history.

Not that weapons were everything. The humans' forms of hunting required extremely accurate perceptions of what was going on around them. Mammals, including humans, have disproportionately large brains that rely on intricate social relationships. So do carnivores, which hunt opportunistically. Brain size is largest among species that hunt large mammals opportunistically while cooperating with and depending on one another. Brain size also correlates with time spent as a juvenile, which in turn relates to exploration, learning, and play. Complex social organization such as that possessed by *Homo ergaster* required intelligence gathering, analysis of that information, and creative uses of it.

These hunting skills, and the weaponry that went with them, developed in Africa after two million years ago and survived virtually unchanged among premoderns everywhere for almost all of that time, until the late Ice Age, some fifty-five thousand years ago.

THE SAHARA DESERT, 1.8 million years ago. Great flocks of birds gyrate above the shallow blue water, set in the midst of stunted grassland that undulates toward the sun-baked horizon. A gentle wind riffles the surface of the lake, where crocodiles lurk near water's edge. As the shadows lengthen, great herds of antelope congregate there, drinking their fill as huge buffalo wallow in brown mud nearby. The plain shimmers in the late-afternoon heat, dark green and yellow after a recent shower. Lions yawn and stretch in the shade of a large boulder, lazily watching the lakeshore for a stray animal, for an easy kill. By another rocky outcrop, a small band of humans sit watchful, long wooden spears close to hand, ready for the evening hunt. On the far horizon, black clouds mass high above the distant mountains, whose sharp peaks stand out in front of the setting sun. As darkness falls, the wary hunters settle down on a high boulder, where lions will never catch them.

The routine of life has not changed for generations—the unending cycle of dawn, midday, and dusk, of colder and warmer months, of dry season and wet. Like the bestiary of which the species is part, *Homo ergaster* shifts backward and forward across the at-this-time better-watered Saharan landscape. And when the desert dries up, and even before, tiny numbers of humans move out to its margins with the animals upon which they prey. Some return southward to the tropical savanna. Others move north and eastward, along the Nile Valley and Red Sea, and the Mediterranean coast. Their slow movements will change history.

Around 1.8 million years ago—the date is still poorly defined—some groups crossed the Sahara Desert into western Asia. Human hunters were not just Africans. They radiated out of the continent as part of much larger mammalian communities that were to colonize Asia and Europe during brief periods of warmer climatic conditions in the north. Unfortunately, the finer details of climate change during most of the Ice Age are still little understood, despite generations of research. But we owe what we do know to the single-minded research of, of all people, a Serbian mathematician.

IN 1912, TWO impecunious young men, the one a poet, the other an engineer, sat in a coffeehouse in Belgrade.[4] They were celebrating the publication of a volume of patriotic verse written by the youthful poet, not with wine but with coffee, for they could afford nothing else. A well-dressed gentleman asked if he could share their table and glanced through the book of poems. A fervent patriot and a bank director, he was so moved by the poems that he ordered ten copies of the book and paid for them on the spot. The jubilant friends turned to red wine instead. By the third bottle, their minds were soaring and seeking new intellectual fields to conquer. "I want to describe our entire society, our country, and our soul," cried the poet. The engineer, Milutin Milankovitch, was not to be upstaged. He announced that he was attracted to infinity. "I want to do more than you. I want to grasp the entire universe and spread light into its farthest corners." After sealing their ambitions with a fourth bottle, the two friends parted.

The wine may or may not have been responsible for the flash of inspiration that afternoon, but the engineer had found a life-changing challenge. Milankovitch was to devote his career to the development of a mathematical theory capable of describing the climates of earth and other planets today and in the past. For thirty years, he labored on a cyclical theory of global climate.[5] He calculated that there were at least two main factors involved in the ebb and flow of the earth's ice caps. The first was fluctuations in the shape of the earth's orbit, from more circular to more oval variations in the tilt of the globe's axis of rotation, and the second was changes in the time of year when the earth was closest to the sun. These factors ran through cycles about every ninety-five thousand, forty-two thousand, and twenty-one thousand years. When everything signaled in the same direction, the earth's climate swung to an extreme of either cold (glacial) or warm (interglacial) conditions.

The Milankovitch cycles generated furious controversy until the late 1960s, when core borings provided long and nearly continuous sequences of microorganism-rich seabed deposits that could be used to reconstruct past climatic changes and especially the size of ice sheets on land. The core record vindicated Milankovitch and his cycles, which provide a general framework for the dizzying climatic oscillations of the past seven hundred thousand years—much of the time humans have lived in Europe. Milankovitch's longest, ninety-five-thousand-year cycle has dominated the past three quarters of a million years, producing a major glaciation every one hundred thousand years. Interglacials, such as the brief one we live in, are the climatic exceptions rather than the rule. About 90 percent of the past five hundred thousand years have been colder than today, and the world's climate has been in transition from cold to warm or back again for about three quarters of that time.

How did the Ice Age affect world geography? Constant fluctuations in the Pleistocene ice caps not only affected their immediate surroundings but also caused global sea levels to rise and fall as much as three hundred feet (ninety-one meters) because of the amount of water locked up in ice sheets and glaciers. The much lower sea levels of glacial periods changed world geography. England was part of the European continent. A land bridge connected Alaska and Siberia. Continental shelves extended far offshore from modern coastlines in parts of North America, Southeast Asia, and elsewhere. At the same time, glaciations in high latitudes reduced the amount of water circulating in the atmosphere, which often led to the spread of deserts in tropical regions.

The most extreme climate changes affected western Europe, where extensive ice sheets often mantled Scandinavia, the Alps, and the Pyrenees. Then, as now, North Atlantic ice affected the Gulf Stream, which carries subtropical water from the western mid-Atlantic toward Europe. The southernmost extent of true glacial waters, the so-called Polar Front, has lain far northwest of Britain for the past 10,000 years, as it did in earlier interglacials. Some 120,000 years ago, during the last interglacial before the late Ice Age, the Gulf Stream flowed around Britain and close to western Europe as it does today, creating climatic conditions much like those of today, and perhaps somewhat warmer. Some 20,000 years ago, Europe was locked in extreme cold, at the height of the last glaciation. The Polar Front now lay across the Atlantic, at about the latitude of Spain. Iceberg-strewn waters lapped northwestern Europe. English anthropologist Chris Stringer believes that polar bears may have swum in the River Thames.

The changes recorded in deep sea borings and ice cores taken from

How do we study Ice Age climate?

Gunz, Mindel, Riss, and Würm: until surprisingly recently, little more than deposits left by advancing and retreating ice sheets defined the Ice Age, or Pleistocene, the last geological epoch.[6] Swiss fish-fossil expert turned geologist Louis Agassiz achieved international fame by identifying it in general terms, but details were long in coming. In 1909, two German scientists, Albrecht Penck and Eduard Brückner, published a history of the Ice Age in which they identified four glaciations, naming them after Alpine river valleys. They estimated that the Pleistocene had lasted about 650,000 years. Gunz and its glacial brothers passed into college classrooms and remained there for generations; they were still taught at Cambridge University when I was an undergraduate in the late 1950s.

The Penck and Brückner scheme soon proved too simplistic, as glaciologists deciphered the complex advances and retreats of North American and Scandinavian ice sheets, did further work in the Alps, and studied raised beaches left by much higher sea levels during the interglacials. Even as recently as a half century ago, most expert opinion dated human origins to about two hundred thousand years before the present. Then Louis and Mary Leakey found two-million-year-old human ancestors at Olduvai Gorge, in East Africa, in 1959–60. The Olduvai dates came from the newly developed potassium argon method, a counting technique that uses radioactivity to measure the ratio of potassium to argon in volcanic rocks, such as abound in the gorge. Since then, potassium argon dating has pushed back the appearance of the earliest toolmaking hominins (formerly called hominids) to at least two and a half million years ago. Soon after the Leakey discoveries, the newly developed molecular clock dated the split between chimpanzees and the ancestral human line to about five million years ago. Human ancestry has much deeper foundations in the past than we once imagined.

When Penck and Brückner studied the Alps, paleoclimatology (the study of ancient climate) was in its infancy and mainly relied on glacial geology. Today's portrait of the Ice Age is far more nuanced and complex, based on a broad range of highly specialized data. A profound revolution in paleoclimatology has taken hold over the past half century, providing the climatological framework for our story.

Deep-sea cores are an important source of knowledge about long-term climatic trends. Hundreds of them now exist, drawn from all the world's oceans. The minute foraminifera in them yield information on changing temperatures over the past three quarters of a million years or more. We now know that there were not four glaciations but at least nine or ten of them, separated by shorter warmer intervals. The world's climate has been in transition from cold to warm and back for at least three quarters of the past 780,000 years, before which time the earth's magnetic field abruptly reversed. Minute benthic (deepwater) foraminifera in sea cores reveal changes in the oxygen isotope ratio (the amount of the heavier ^{18}O in relation to the lighter ^{16}O) in the ocean over time, which are a reflection of global-scale ocean chemistry changes as ice built up on the continents during glaciations. Variations in the oxygen isotope ratio gave rise to what are now termed *marine isotope stages (MISs)* or *oxygen isotope stages (OISs)*, used as worldwide climatic stages. We are currently in isotope stage 1, and the last glaciation, when the world's oceans were more enriched with the heavy isotope, is represented by OIS 2 to 4 (see figure 3.3 for details).

Ice cores drilled out of the Antarctic and Greenland sheets, mountain glaciers in the Andes, the Tibetan Plateau, and elsewhere are now providing an increasingly precise record of changing climate and shifting temperatures far back into the past. Ice cores from Antarctica now extend as far back as about 800,000 years ago, with a record of identifiable glacials and interglacials going back to more than 650,000 years ago.

Inevitably, the climatic record is somewhat generalized until the last relatively warm interglacial, which began about 128,000 years ago. We know about most of the more prolonged temperature swings but lack information on minor changes, such as warm periods that lasted a couple of thousand years or so. The last glaciation is documented in much greater detail, to the point that we now know that there were two intensely cold periods, one about 70,000 years ago, when the Neanderthals inhabited Europe, the other about 21,500 to 18,000 years before the present, the Last Glacial Maximum.

For all the new information, all we really have is a set of snapshots of glacial climate from all kinds of what are known as proxies, indirect records of climatic change: a wide variety of archives from which temperature and rainfall have to be extrapolated. These include ice-core curves, growth records from tropical coral, and local records of vegetational changes derived from minute pollen grains from ancient trees and plants preserved in marshes and swamps. All these chronicles are inevitably incomplete, forming a patchwork of information from different sources, which can include the annual growth rings of trees, and also data from mollusca, rodent bones, changing sea levels, and even earthworms—to mention only a few.

The search for Ice Age climate has still hardly begun.

Greenland, Antarctica, and elsewhere, even strung out over thousands of years, are even more impressive when one realizes that some changes from glacial to interglacial occurred not over a period of millennia but over time measured in centuries, even decades, perhaps sometimes less than a human generation. Milankovitch's cycles are still at work, despite an inexorably warming world. Theoretically, we could be plunged abruptly into another episode of the intense cold that has dominated so much of the past three quarters of a million years. Just when is a matter of debate, but perhaps in ten thousand years or so, if Milankovitch was correct, provided that humanly caused global warming does not upset the natural cycles of cold and warm. And when the cooling does occur, it may well be rapid enough to change people's lives within a generation or so.

Interglacials were key climatic players in Europe's early history. Brief warmer cycles effectively opened a gate between tropical Africa and other parts of the world. For tens of thousands of years, cold conditions in the north reduced water circulation in the atmosphere and caused very dry conditions in the always-arid Sahara Desert. However, when ice sheets shrank and the Gulf Stream warmed Europe, slightly wetter conditions brought shallow lakes and extensive tracts of semiarid grassland to the Sahara. Some large watercourses fed by monsoon rains flowed across the desert from the Saharan highlands, forming now-vanished natural highways for animals and humans, which can be mapped only by satellite radar. The desert acted like a giant set of lungs, sucking animals, including humans, onto its grasslands at times when the climate north of the Mediterranean was warm, perhaps sometimes even warmer than today.

One of these brief—but poorly researched—more temperate periods brought *Homo ergaster* and other mammals out of Africa and, ultimately, into Europe. Not that this was a deliberate migration, for human societies lived by the rhythms of familiar routines that governed their daily existences. On an annual basis, *Homo ergaster* must have covered relatively short distances, maintaining close contact with other members of the social group, and perhaps neighboring bands. Everyone was accustomed to climatic shifts, to droughts and periods of higher rainfall. The boundaries of where they could settle were set by their ability to reproduce and the availability of food and water. And, in the fullness of time, *Homo ergaster* (usually called *Homo erectus* outside of Africa) settled in western Asia and then in Europe (figure 2.2).

MOVING INTO EUROPE presented a challenge for humans, who were still basically tropical animals. Days were shorter for much of the year, winters were severe even during interglacials, and the seasons were very marked. Plant foods were in much shorter supply, which meant that game assumed even greater importance in human diet. One could not survive without an impressive expertise in big-game hunting.

Where and when, then, did premoderns first enter Europe? Most likely, the first Europeans entered their new homeland from western

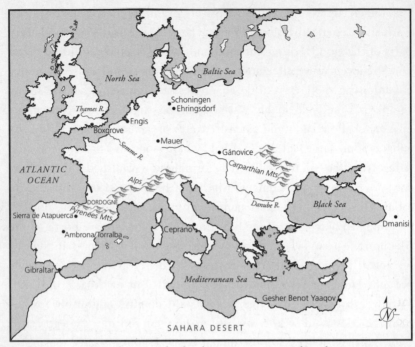

Figure 2.2 *Map of sites and other locations mentioned in chapters 1 to 4.*

Asia across what is now Turkey, just as *Homo sapiens* would hundreds of thousands of years later. They lacked the watercraft to cross open water such as the Strait of Gibraltar. The immigrants were very thin on the ground, so it is hardly surprising that we know almost nothing about them. Some of them visited a small lake at Dmanisi, in Georgia's Caucasus Mountains, during a warm interval about 1.75 million years ago.[7] Here the bones of *Homo ergaster*–like individuals and crude stone chopping tools were found under a medieval village. Unlike their African relatives, some of these people were short in stature. Their skulls display considerable anatomical variation, including traces of more primitive traits that predate *ergaster* in Africa. Interestingly, one individual had lost all but one of his teeth long before his death, for layers of bone fill his teeth sockets. He could only have consumed soft food that could be swallowed without chewing, so other members of the band must have looked after him.

Dmanisi has strong African associations. The anatomy of the humans reflects their African ancestry. Fragmentary animal bones excavated from the site include those of short-necked giraffes and ostriches, both animals of African origin. The people were part of a large bestiary, which also included such dangerous predators as saber-toothed cats. To survive must have required fleetness of foot and impressive hunting skills.

Tiny numbers of human immigrants penetrated Europe during these millennia, so few that their bones rarely survive. When they do, the bones are often too fragmentary to tell us much. In 2008, Spanish researchers discovered a 1.1- to 1.2-million-year-old fragmentary human jaw, as well as animal bones and stone tools, in the Sima del Elefante cave, in the Sierra de Atapuerca region of northern Spain, far to the west.[8] We know that people were living at Ceprano, in central Italy, about 800,000 years ago, for a human skull found there resembles those of fossil humans from the Gran Dolina cave in the Sierra de Atapuerca dating to before 780,000 years ago. A few humans were in Britain by 700,000 years ago.[9]

A palimpsest of human bones, scatters of stone artifacts, and animal bones tell us virtually nothing about the first Europeans. One fascinating question immediately arises. Did they use fire? Could one survive in what were soon to become much colder environments without it? We don't know. Theories about the origins of fire abound, some of them placing its taming in tropical Africa at least 1.8 million years ago, but the evidence is little more than some fire-damaged stone that could have resulted from a wildfire. People may well have used fire from such sources thousands of years before they were able to light and control it, essential skills if it was to be used for cooking, protection, and warmth in cold climates.

The earliest known domesticated fire was made at a 790,000-year-old campsite at Gesher Benot Ya'aqov, in the Jordan Valley.[10] Small numbers of people had already settled in temperate environments by this time, apparently without fire. But hearths lit at will provided warmth and, perhaps even more important, protection against predators, both in the open and in caves and rock shelters, so we can assume that fire was a decisive innovation as far as European settlement was concerned. Even with fire, however, much of glaciated Europe, with its savage winters, may have been too cold for human settlement. Like other mammals, the few groups in the north may have retreated southward into warmer areas during colder cycles. It may be no coincidence that the earliest Europeans in central and western Europe appeared in Spain and Italy and that half a million years were to pass before human populations increased significantly during another period of warmer climate.

THE DMANISI SKULLS from Georgia display features reminiscent of their African ancestor, *Homo ergaster*, but was *ergaster* the direct ancestor of the Neanderthals? In 1907, workers in a sandpit at Mauer, near Heidelberg, Germany, unearthed a thick, chinless human jaw. Local experts promptly called it *Homo heidelbergensis*, "Heidelberg Man." Few of their colleagues took the label seriously. They considered the incomplete fossil a European version of *Homo erectus* dating to about five hundred thousand years ago. Then, over the next three quarters of a century, a handful of fossils from Africa and throughout Europe revealed people with significant differences from *Homo erectus* who looked more like *heidelbergensis*. All had a higher, more filled-out braincase, reflecting a larger brain size closer to that of modern humans, a face that was reduced compared with the jutting countenances of earlier humans, and considerably thinner bony reinforcements of the skull. It turns out that the German anthropologists of a century ago were correct. *Homo heidelbergensis* was indeed a distinct human form with some more advanced features than *erectus*, but with more primitive anatomical characteristics than those of the Neanderthals and modern humans.[11]

Like *Homo ergaster*, *heidelbergensis* was African, and probably evolved there from earlier human stock about six hundred thousand years ago. Just when and how the newcomer moved out of Africa and into Europe remains a mystery, but the process must have been very similar to that of earlier humans—a natural movement alongside other mammals. Thanks to recent finds in Spain's Sierra de Atapuerca, we know that this little-known human was the direct ancestor of the Neanderthals.

Sima de los Huesos, "the pit of the bones," is a small, muddy chamber in the sierra at the bottom of a forty-three-foot (thirteen-meter) vertical shaft, reached by walking, crawling, and using ropes and a metal ladder. So far, the pit has yielded over two thousand jumbled human fragments from at least thirty-two men, women, and children. No one ever lived in this inaccessible chamber, and why human bodies were dumped here is a mystery. A random combination of ancestral and more modern features appears throughout the Sima bones. The most complete skull has a lower jaw that looks like a scaled-down version of the Mauer mandible. Meanwhile, the face of the same individual displays a projection around the nose characteristic of Neanderthals. These were powerfully built, very strong people who displayed considerable anatomical variation. The males stood about 5 feet 6 inches (1.7 meters) tall and weighed about 136 pounds (62 kilograms), the women being somewhat shorter and lighter. Their receding foreheads displayed prominent browridges, and their jaws were massive and chinless, all features that survived in Neanderthal anatomy. Spanish scientists classify the Sima finds as a late form of *Homo heidelbergensis*. Others, such as anthropologist Chris Stringer, believe they are early Neanderthals. But whatever the classification, the bones show that an evolutionary transition was under way in European human populations of about four hundred thousand years ago. These changes would lead some two hundred thousand years later to the Neanderthals. Amphibian, lizard, and snake remains tell us the mean annual temperature was slightly warmer than today, averaging between fifty and fifty-five degrees Fahrenheit (ten and thirteen degrees Celsius).[12]

Exactly when the first true Neanderthals appeared remains unknown, for human fossils dating to between 250,000 and 70,000 years ago are scant. There are tantalizing hints: A Neanderthal-like skull from Ehringsdorf, near Weimar in central Germany, dates to perhaps about 200,000 years ago. As the climate grew colder around 100,000 years ago, some Neanderthals camped around a mineral spring at Gánovce, near Poprad in Slovakia. At the time, coniferous trees were spreading into a region long covered with oak forests. Thereafter, Neanderthal remains become more common as intense cold settled over Europe and Eurasia.

HOMO HEIDELBERGENSIS REMAINS a shadowy presence: how did these people live, beyond apparently possessing fire? We can only generalize from a handful of well-documented sites.

By half a million years ago, a few thousand hunter-gatherers dwelled throughout Europe, based for the most part in river valleys and other places where animals and plants abounded. Much of the time, they lived in small bands, with only sporadic contact with neighbors. Most people would have met only thirty to fifty people, if that many, during their brief lifetimes. The chances of hunting accidents or falling victim to predators were high for people possessing simple weaponry and living in an environment rife with large, dangerous animals and menacing predators like lions.

For all that hazard, *Homo heidelbergensis* groups survived and flourished. For thousands of years, they lived by the banks of large rivers like the Somme, in northern France, and the Thames, preying on the game that populated these well-watered, sheltered locales, during millennia when climatic conditions were as warm as, if not warmer than, today. Thousands of their finely made stone hand axes lie in the gravel of European rivers, most once used for butchering game (figure 2.3). Early archaeologists collected these axes by the sackful, some of them magnificent examples of the stoneworker's art—long, with lanceolate points and finely chipped bases, so fine, in fact, that you wonder if some of them had ceremonial purposes or were even made just for display.

Stone axes and flakes tell us little about the people who made them, or the way they lived. Fortunately, we sometimes come across a prehistoric event frozen in time, where the preservation is so good that it is almost as if one were there witnessing it. The German archaeologist Hartmut Thieme experienced such a moment when he was called in to excavate Ice Age levels exposed during open-pit coal mining at Schöningen, in northern Germany.[13] He recovered the remains of a successful hunt—and the weapons used in the chase. One summer about four hundred thousand years ago, a group waited for game on the shores of a shallow, elongate lake. The area was dry and treeless, so they could watch easily for potential prey. Their patience was rewarded. They stalked, intercepted, killed, and butchered a herd of small wild horses.

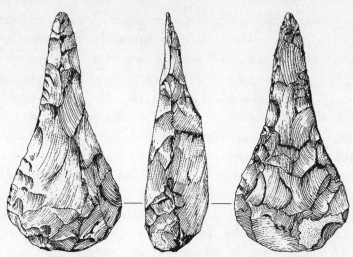

Figure 2.3 *Acheulian hand axe from the Thames Valley, England. The exact age is unknown but it is over two hundred thousand years old. The label Acheulian comes from the northern French town of St. Acheul, where such tools were first excavated in the mid-nineteenth century.*

Twenty-five thousand bone fragments document the hunt and the butchery that followed, carried out with sharp stone knives and scrapers that the people had brought to the kill. There are no signs of stoneworking on-site.

The site lay in an organic mud deposit, which offered preservation conditions so perfect that Thieme recovered not only the bones of the butchered horses and the stone tools used to dismember them but also the long wooden spears used in the chase.[14] Each was between 6 and 8 feet (1.8 and 2.4 meters) long, made from a single, carefully trimmed sapling. The hardest wood at the base formed the point, the maximum thickness and weight lying about a third of the length back from the tip. A modern replica hurled under controlled conditions displayed good ballistic qualities, with impressive penetrating powers even at some distance. The hunters may also have taken geese at the lakeside using throwing sticks, a single example of which came from the excavation, identical in shape and size to those used hundreds of thousands of years later by Australian Aborigines.

The group (or groups) occupied a large hunting camp complete with a row of hearths about three feet (one meter) apart. Nearby lay large bison bones, which bore traces of repeated cut marks, as if the people were slicing meat into strips for drying. The hearths lay on higher, dry ground. Perhaps the camp was in use in late summer or fall, when water levels were low and rainfall sparse. Then in late autumn, the first snow covered the bones and abandoned tools under a thick layer of decaying reeds before rising waters in spring preserved the site for posterity. Schöningen hunting weaponry had progressed far from the simple pointed sticks of earlier times.

At about the same time, roughly 370,000 years ago, another group of people occupied three simple tents in a summer camp at Bilzingsleben, in central Germany.[15] Here the inhabitants hunted red deer in the surrounding oak forests. They also scavenged the carcasses of elephants and rhinoceroses, took the massive bones, smashed them across wooden and stone anvils, then chipped them like stone tools to make robust edges for scraping skins and other tasks, treating bone just like stone.

Hunting with nothing more than long wooden spears and throwing sticks was never easy, even for the most experienced hunter. Lucky was the band that managed to come across predator kills or large animals mired in swamps, like the men and women who butchered elephants in a swampy valley at Ambrona, in central Spain, about three hundred thousand years ago. At nearby Torralba, most of the left side of a large elephant was cut into pieces. In both cases, the hunters broke open the massive skulls to get at the brains. Hand axes and other butchering tools littered the sites.[16]

We can imagine the busy scene, as gore-covered men and women clamber over the massive beasts, peeling back the thick hides with sharp-edged stone cleavers, hacking through joints and along ribs to dismember the carcasses, cutting strips from choice cuts to be carried away and dried in the sun and wind. Ever watchful, men with spears keep a close lookout for lions, while hyenas hover close at hand. Come twilight, the band moves away to the protection of hearth and camp, and the waiting predators move in for their turn.

In many places, people probably lived in relatively large groups of perhaps as many as thirty people at times, both to reduce the danger from carnivores and to improve the chances of taking larger animals. Some of their hunting activities involved returning to the same place for generations. At Boxgrove, in southern England, a three-hundred-foot (ninety-meter) cliff overlooked a shallow lake fed by a spring.[17] Here hunters lay in wait near the water on many occasions, killing and butchering animals such as bison, deer, horses, and rhinoceroses. One horse shoulder blade displayed a hole made by a wooden spear about two inches (fifty millimeters) across. A forensic pathologist conjectures that it was spinning when it made the wound, as if the weapon resembled the long Schöningen spears. The meat yields must have been enormous, with 1,550 pounds (700 kilograms) of flesh coming from a single rhinoceros carcass. Much of the meat must have been dried, as it was at Schöningen.

The staple at Schöningen and elsewhere may have been hunting, but plant foods were also important, especially when fruits or nuts were in season and easily obtained by small foraging parties.

EVEN AS EARLY as four hundred thousand years ago, the basic dynamics of what was to become Neanderthal life were long in place. Human existence revolved around the seasons, the ripening of nuts and seeds, the migrations of animals large and small. In some places, such as Italy, people also collected mollusks from the seashore. From spring through fall, the bands moved through their hunting territories, foraging for edible plants, collecting toolmaking stone, preying on animals when the opportunity arose. These were months when people stockpiled sun-dried meat for the long months of winter, when everyone stayed close to home. In winter, each band settled in a sheltered location, where game could sometimes be taken and fish could even be lured by hand from still river pools.

For more than two hundred thousand years, the ancestors of the Neanderthals followed a simple way of life that was flexible, an existence that bent with the climatic winds and adapted readily to extreme cold or prolonged warmth. They lived in familiar environments, even when major climatic changes caused them to shift southward, into more temperate surroundings. Judging from slowly changing styles of stone axes, innovation was rare and technological change almost imperceptible. The rhythm of daily life varied little from one generation to the next, just as the lives of animals followed predictable and familiar paths of migration and dispersal, life and death. Humans were collaborative predators among predators, both hunters and the hunted, effective at survival thanks to their expertise with wooden spears, their stalking ability, and their painfully acquired knowledge of animals and plants. And, over two hundred millennia, they gradually evolved into the Neanderthals, the primordial Europeans encountered by the Cro-Magnons.

CHAPTER 3

Neanderthals and Their World

FEEL SORRY FOR THE NEANDERTHALS. Their name still evokes persistent images of hairy cave people enamored of brute force. To be described as "Neanderthal"—and I was myself so labeled on occasion in my youth—implies one is guilty of boorish and antisocial behavior. Such grossly inaccurate stereotypes have haunted these agile and versatile people ever since their discovery. Even today, we are ambivalent about the Neanderthals, perhaps because we are uneasy about accepting them as close relatives.

The first Neanderthal fossil, the skull of a child aged two to three years, came from the Engis cave, in Belgium, in 1830. Even at such a young age, it displayed incipient browridges over its eye sockets. In 1848, a cave on Gibraltar, at the southern tip of Spain, yielded a nearly complete Neanderthal skull with beetling brows. Neither of these finds attracted much attention. The Gibraltar cranium was merely thought to be "ancient." Nearly twenty years passed before it was identified as that of a Neanderthal. The Engis child remained an anatomical mystery for more than a century.[1]

In 1856, the owner of a limestone quarry found some large bones in clay deposits shoveled out of caves in the cliffs of the Neander Valley, near Düsseldorf in northwestern Germany. The workers had thrown the clay sixty feet (twenty meters) down the slope. Thinking that the bones came from a cave bear, he showed them to a local schoolteacher and amateur naturalist, Johann Carl Fuhlrott. Fuhlrott realized at once that the bones were human and passed them on to a well-respected anatomist, Hermann Schaafhausen. The judicious Schaafhausen studied the bones carefully and presented his findings in 1857, two years before Charles Darwin published *On the Origin of Species*. He proclaimed the bones to be the remains of a barbarous member of a "very ancient human race."

Schaafhausen's findings came at a time when most people still believed in the literal historical truth of the Scriptures. Church chronologies proclaimed that the world had been created in 4004 B.C., giving less than six thousand years for all of human existence. The Neander discovery was immediately controversial. Schaafhausen's critics weighed in with all manner of theories. Some said that the bones were those of a Mongolian Cossack, one of the soldiers who had chased Napoléon across the Rhine in 1814. Rudolf Virchow, the most eminent anatomist of the day, dismissed the fossils as the bones of a recently deceased pathological idiot, who had suffered from rickets and lived as a hermit in one of the caves.

The critical moment came seven years after the discovery. English biologist Thomas Henry Huxley, a fervent supporter of Darwin's theories of evolution and natural selection, carried out an exhaustive study of the Neander finds. In his *Man's Place in Nature* (1863), a classic of early anthropology, he declared that the bones, which had some apelike associations, were those of an ancient human being, an ancestor of *Homo sapiens*. Huxley's reasoned discourse on the Neanderthal skull started a controversy that still rages today: what was the relationship between these archaic people and their anatomically modern successors—ourselves? He boldly traced human ancestry back to our closest living relatives, the chimpanzee and the gorilla.

Man's Place in Nature sent a profound frisson through devout Victorian society, where the Church of England still wielded enormous power and preached the literal historical truth of the biblical Creation. To be accused of heresy, to challenge Genesis, 1, was to commit a serious offense and risk social exile.[2] Respectable churchgoers were appalled. "Let us hope it is not true," cried one horrified lady upon hearing that humans were descended from apes, "but if it is, let us pray it will not become generally known." Huxley and his followers were tireless in their defense of Darwin's theories, although the great man himself was circumspect about human origins, beyond pointing to Africa, with its diverse ape populations, as a possible cradle of humankind. In time, science prevailed over dogma, largely because more Neanderthal fossils came to light elsewhere— in Belgium, France, Germany, and Spain. The new finds proved that the Neanderthal skull was no isolated aberration, that beetle-browed Neanderthals had once flourished over a wide area of Europe. With their protruding jaws and heavy browridges, the Neanderthals were quite unlike the rounded-head *Homo sapiens* whose ancient remains had been discovered previously, the ancestors of modern humans. The Neanderthals came to light at a time when the first anthropologists were describing all manner of exotic, non-Western cultures, some of them hunters and gatherers with only a handful of simple artifacts. A vast intellectual and social chasm separated such "savage" societies from the dazzling complexity of Victorian industrial civilization. In an era when doctrines of racial superiority reigned supreme, the Neanderthals looked "primitive" and, inevitably, acquired a reputation for brutish savagery.

In 1908, a complete male Neanderthal burial came to light in La-Chapelle-aux-Saints rock shelter, in the Vézère Valley of southwestern France. Marcellin Boule, the leading French biological anthropologist of the day, pored over the bones. Inadvertently, he developed a grossly misleading caricature of the Neanderthals in the process. Unfortunately, he missed the chronic arthritis that had crippled the man's spine, which had caused this particular Neanderthal to stoop. Boule described a slouched, stooping hunter with a large, primitive head poked forward on a short, thick neck, the epitome of a hulking, dim-witted man who lacked the well-developed face and high forehead of modern humans. Hardly surprising, he declared that Neanderthals were an evolutionary dead end and not direct ancestors of *Homo sapiens*.

For all its appeal to cartoonists drawing cave people, Boule's characterization wasn't, by any means, accepted by everyone. In the 1930s, the Harvard anthropologist Carleton Coon produced a drawing of a Neanderthal dressed in modern clothing. He pointed out that a Neanderthal could ride on the New York subway without drawing much attention, provided he was dressed properly and shaved.

One can hardly blame Boule for his misleading portrait. He worked at a time when there were far fewer Neanderthal remains to compare.

Today's Neanderthal scholars have access to some five hundred finds, including the more or less complete skeletons of some twenty men, women, and children. They also use a broad array of sciences and methods that were unknown in Boule's day, among them genetics, population studies, and sophisticated approaches to skeletal anatomy. These newer studies have shown that the La-Chapelle man is an anomaly, that in fact the Neanderthals were nimble and strong, fleet of foot, and capable of surviving a remarkably broad range of climatic conditions and meteorological shifts. They were clever people, with considerable intellectual abilities—but they were not fully modern humans.

I ONCE WAS lucky enough to handle a Neanderthal cranium. The skull-cap, about the weight of an avocado, lay in a carefully padded wooden drawer in a private museum near Les Eyzies. I picked it up rather casually, under the impression that it was a modern specimen, and nearly dropped it when my host informed me that it was the skull of a Neanderthal from the nearby La Chaise rock shelter. As I turned the smooth, brown object around in my hands, I felt a curious affinity for its owner, who had grown up, hunted, and died close to where his skull had now ended up. I've been fascinated by the Neanderthals ever since.

Inevitably, most of what we know about them comes from generations of intense measurement, CT scans, genetic testing, and study of specimens like the cranium from La Chaise, complicated by small samples and diversity among Neanderthal populations. What can we say about them from this research? They were striking people, with brains as large as our own, but their heads were shaped differently. We have high foreheads and rounded heads, whereas the Neanderthals had long, low skull vaults, which were larger and protruded at the back. A pair of large, rounded, and continuous ridges overhung the eye sockets. The front of the skull was somewhat flattened and constricted like that of much earlier, archaic humans. This may be significant, for it is in this area, the so-called prefrontal association cortex, that much of our thinking takes place. In other respects, casts of the interiors of Neanderthal brain cases reveal a substantially modern level of external brain organization and asymmetry.

Neanderthal faces protruded markedly, the cheekbones sweeping sharply backward from a very large nasal aperture, whereas we moderns tend to be smooth browed, with small faces tucked below the front of the skull. Our smaller faces are also reflected in our jutting chins and the short tooth rows of our lower jaws. In dramatic contrast, the Neanderthals had long rows of teeth and lower jaws that either receded or were vertical. Their front teeth were large and often endured severe wear, because they pulled skin, food, or fibers across them. DNA tests on forty-three-thousand-year-old Neanderthals from El Sidrón cave, in northern Spain, revealed a pigmentation MICR gene that hints that some Neanderthals may have had pale skin, red hair, and even, perhaps, freckles, possibly an adaptation to let in more sunlight to manufacture vitamin D in northern environments.[3]

They were robustly built people for the most part, often shorter in stature than ourselves, aptly described by several writers as "fireplugs," standing about 5 feet 5 inches (1.6 meters) tall and weighing about 185 pounds (84 kilograms) Their limb proportions differed from those of *Homo sapiens*, with relatively short upper-arm and leg segments. They had large joints and thick-walled arm and leg bones. A glance at the structure of their shoulder blades reveals very powerful upper-arm musculature, while their hip bones hint at some minor differences in gait compared with us. Such differences did not prevent them from running fast and moving nimbly. By seventy thousand years ago, European Neanderthals had adapted to the climate and physical rigors of a bitterly cold world, where temperatures oscillated dramatically. Their bulky shapes conserved heat by minimizing the surface area of skin exposed to the cold. The large internal volume of their prominent noses may have served to warm and moisten the cold, dry air they inhaled.

For all their bulk and compact bodies, the Neanderthals were agile, tough people, who may have reached puberty somewhat earlier than modern humans—the data on this point is still uncertain. Their skeletons bear witness to the hard lives they lived, to healed broken bones and other injuries, some of considerable severity. Many of these injuries must have resulted from close encounters with very dangerous animals such as bison or aurochs, with their huge, razor-sharp horns. More than one scholar has compared their injuries to those suffered by modern-day rodeo riders upon landing after a fall. To deliver a fatal spear thrust often meant that the hunter literally had to jump astride his prey. The Neanderthals' thick bones and powerful muscles combined the physique of a wrestler with the endurance of a marathon runner (figure 3.1).

These, then, were the people who survived in Europe and in the Mediterranean world for an immensely long time: from before two hundred thousand years ago until about thirty thousand years before the present. DNA sequence fragments from twelve fossils divide the Neanderthals into three, maybe four, subgroups: one in western Europe, another in southern Europe near the Mediterranean, a third in eastern Europe and the Near East, and a fourth that may have thrived in western Asia.

They lived in a profoundly challenging world. Great mountain ranges rising to altitudes of sixty-five hundred to nearly ten thousand feet (two thousand to three thousand meters) above sea level shielded

Figure 3.1 *Neanderthal hunters. Giovanni Caselli.*

the Mediterranean zone climatically from colder environments. North of the mountain ranges, the hilly forelands of western France, southern Germany, and further eastward gave way to the North European Plain, which began in the dry bed of the North Sea during the low sea levels of the late Ice Age. The plain gradually widened eastward toward Russia and the Ukraine. We can trace Neanderthal settlement over the continent from archaeological sites by the presence of their distinctive stone tools. Most of these camps lay in the coastal zones of Spain and Italy and along the Atlantic shores of Portugal, northern Spain, and France (figure 3.2). A wide belt of less densely concentrated sites straddles the fiftieth parallel from Britain to central Russia, but there are few signs of occupation further north during the last glaciation. Significant numbers of Neanderthals lived in the Crimea and along the Black Sea.

The earliest known Neanderthals lived through a long cold period, then through the warm millennia of the last interglacial, between

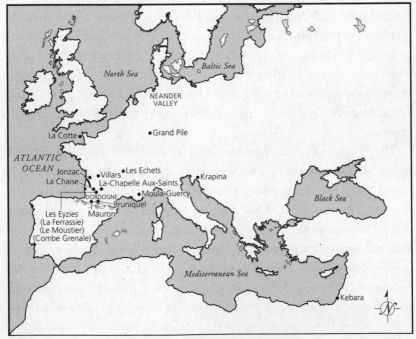

Figure 3.2 *The Neanderthal world.*

about 128,000 and 115,000 years ago, then through much of the last glacial period and its many climatic swings. Their ability to survive for such a long period of time went far beyond their biological adaptations. No other human beings had mastered such a broad sweep of environments—nor would they until the Cro-Magnons came along. That the Neanderthals did so with the help of fire, crude skin cloaks, and a mere handful of basic tools and weaponry is remarkable. They survived successfully in environments that today's adventurers face with a bewildering array of multilayered clothing and survival technology. They did so using strategies that went virtually unchanged for hundreds of thousands of years.

HAD A NEANDERTHAL family of seventy thousand years ago been transported back a quarter of a million years to join *Homo heidelbergensis*, they would have had no trouble resuming their daily routine. How do we know this? Our impressions of an unchanging lifeway come from informed guesswork and from studying infinitesimal changes in stone tools and hunting methods over tens of thousands of years. Just like their ancestors, Neanderthals danced the dance with their environment. The waltz never ceased, whether in glacial or interglacial times, but its tempo changed throughout the year, driven by the passage of the seasons and longer-term climatic shifts that unfolded over centuries rather than generations.

The rhythm of premodern life remained effectively the same for millennia, despite some minor technological changes. Like their prey, each human band ebbed and flowed across relatively small territories with the seasons. They moved with the animal communities of which they were part. The pace quickened in early spring, when deciduous trees burst into leaf and melting snow swelled rivers and streams. Red deer fed on fresh grass shoots in forest clearings. Aurochs grazed in swampy water meadows. This was the season when reindeer moved from sheltered winter valleys to the open plains, crowding river crossings and narrow passes, where hunters could spear them from nearby cover.

Spring was a time of recovery, of increasing abundance, when the bands fanned outward into more open terrain. By early summer, the growing season was in full swing. Plant foods of all kinds ripened; fruit came into season. During the short summer months, such foods must have briefly assumed great importance in the Neanderthal diet, for they were close at hand and often abundant. While women foraged for plant foods, hunting never ceased. The hunters ranged widely, constantly on the move, watching closely for telltale signs of a predator kill, perhaps of a bison or aurochs. They must have moved carefully, alert for fresh droppings, for flattened grass, the distinctive tracks of prey that had passed through but an hour before. They would have known birdcalls and aurochs bellows, which they would have imitated flawlessly. To hunt successfully and survive, they had to know the habits of their quarry as well as they knew their own kin. Success depended as much on stalking as it did on weapons.

Fall came early in colder times, with frosts by September, even if the days were still warm. Colder days and cold winds brought the reindeer and other migrating animals southward into more sheltered places. Again, the Neanderthals would have preyed on the herds moving through narrow valleys and across streams. By fall, too, the last nuts would have ripened, so the diet would have become almost entirely meat based. The bands moved into valleys and protected places like rock shelters and caves, even if they still moved from place to place. Much of their diet may have come from dried meat set aside from summer hunts.

Now came the lean months, when the snows came and temperatures plunged. No one ventured far from camp during the short days, well aware that predators lurked in the forests in search of prey. As always, hunting was opportunistic: an aged aurochs or bison, a deer painstakingly stalked as it pawed through thin snow looking for grass, a boar drinking from a fast-running stream. Judging from historical records of Inuit people in the Canadian Arctic, people may have gone into simple hibernation, clustered round fires, spending much time huddled up against one another wrapped in thick furs, and sleeping a great deal.[4]

The specifics of the routine, especially the unconscious movements that coincided with the seasons, might change from location to location, but the essential pattern of life, dictated not so much by human decision as by the realities of animal migrations and changing seasons, remained much the same as it had been since the days of *Homo heidelbergensis* over two hundred thousand years earlier. Given the limitations of the Neanderthal mind and of their technology, and the rigors of the late Ice Age climate, it could be no other way.

"THE REINDEER AGE"—the Victorians considered the world of the late Ice Age as a time of extreme cold, of glacial climate so severe that Europe was at, or beyond, the limits of human endurance. They thought of the Neanderthals as short, compact individuals who eked out a hand-to-mouth existence in a world of practically continuous winter. Nothing could be further from the truth. While the Neanderthals did indeed flourish during periods of intense cold, they in fact lived through a glacial period that was marked by constant, and often sudden, climatic shifts.

The heyday of the Neanderthals was the first part of the last glacial period, first identified in the Alps, then in Scandinavia, and named by glaciologists the Würm (Alps) or Weichsel (Scandinavia) glaciation. Perhaps it's better to use the generic term *late Ice Age*, which encompasses a glacial interval that lasted from about 115,000 years ago until 15,000 years before the present, when irregular warming began (figure 3.3).

Generations of climatologists assumed that late Ice Age climate was unremittingly frigid. They painted a forbidding portrait of Europe in a deep freeze. Huge ice sheets mantled Scandinavia and much of Britain, the Alps were completely icebound, and ice sheets covered the Pyrenees Mountains between what is now France and Spain. Conditions appeared so extreme that some archaeologists questioned whether either Neanderthals or Cro-Magnons could have survived them and believed that both may have been driven to the brink of extinction. You only have to glance at temperature changes reflected in Greenland ice cores to realize that the climatic reality was much more complex.[5]

Thanks to the various sources described in the box, we now know that the last interglacial ended about 115,000 years ago, with slow cooling

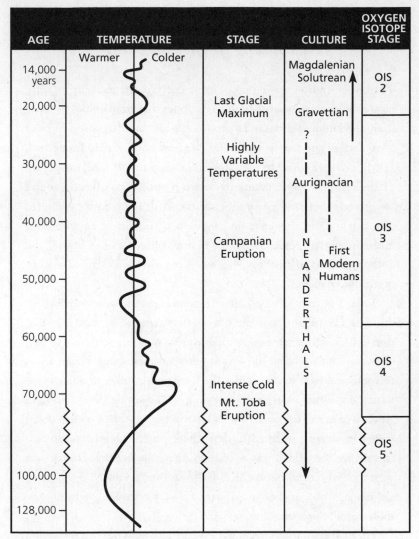

AGE	TEMPERATURE	STAGE	CULTURE	OXYGEN ISOTOPE STAGE

Figure 3.3 *A generalized summary of late Ice Age climatic change, with major cultural developments in Europe.*

interrupted by two major warmer intervals when temperatures returned to near-interglacial levels. Sea levels fell dramatically as major ice sheets accumulated in North America.[6]

Between seventy-four thousand and sixty thousand years ago, the cooling intensified. A huge ice sheet covered much of Scandinavia, but

What was late Ice Age climate like?
Assembling the jigsaw puzzle

The last glaciation was no monolithic deep freeze, but only recently have paleoclimatologists begun to decipher the complexities of more than one hundred thousand years of the climatic backdrop to both Neanderthal and Cro-Magnon life. We now have a broad framework of what is often called oxygen isotope stage 3, or OIS 3 (a long period of the last glaciation, dating to between about sixty thousand and twenty-one thousand years ago; see the box "How do we study Ice Age climate?" in chapter 2 and figure 3.3), thanks to an ambitious multidisciplinary research project that assembled climatologists and anthropologists to piece together what can only be described as a climatic jigsaw puzzle.

Today's portrait is based on high-resolution climatic modeling using data from many sources. We are beginning to acquire information on changing regional temperatures, rainfall, snow cover and snow depth, and the all-important temperature gradients from north to south and west to east across the European continent. Some fascinating details are emerging. For example, the spring thaw arrived in April in eastern Europe but much later in the west, and temperatures were considerably warmer south of the Pyrenees. During warmer intervals, winter rainfall was as high as, if not higher than, today over western and central Europe. Rainfall was much reduced during cold millennia. These are but snapshots of a very complex, and still little understood, European climate.

The high-resolution Greenland ice-core record is the most important source of information about late Ice Age climate, for it reflects climate changes in the North Atlantic and perhaps in the more maritime parts of western Europe. Whether it provides a record of simultaneous climatic shifts in the Mediterranean, central Europe, and eastern Europe is still uncertain, although the shifts seem to be duplicated in the western part of the former. Deep-sea cores provide very similar information.

One important climate engine was the so-called Dansgaard-Oeschger (D/O) oscillations, which occurred at least twenty-five times during the last glaciation. Best documented in the Greenland ice cores, the D/O oscillations were rapidly warming episodes followed by slow cooling over a longer period. Major or minor, they seem to have occurred (and still occur) about every 1,470 years and may have been connected to changes in the North Atlantic circulation perhaps triggered by an influx of freshwater, from ice sheets that became unstable and released huge quantities of freshwater into the ocean. Apparently, the effects were widespread. Stable isotope records from a stalagmite in the Villars cave, in southwest France, document most of the D/O events between 83,000 and 32,000 years ago. The stalagmite shows that temperatures dropped considerably between 75,000 and 67,400 years ago before giving way to generally warmer, fluctuating temperatures. One major warm event took place between about 46,800 and 42,300 years ago, when it ended abruptly, an occurrence also marked by a violent seesaw in the Greenland ice cores.

Pollen analysis, or palynology, is a powerful tool for chronicling vegetational change. The minute pollen grains that are all around us survive in many organic geological deposits and in archaeological sites. They provide local sequences of vegetation change as a basis for reconstructing ancient biomes, using models based on plant functional types (PFTs), which are given scores. These, and actual plant remains, show that both warmer and cooler events supported vegetation that was very different from today's. During warmer intervals, the steppe-tundra may have been a mosaic of parkland with savannalike vegetation that supported many herbivores. The vegetation patterns from pollen data tend to correlate with the climatic shifts in the Greenland ice cores and in deep-sea cores.

The rich bestiary of the last glaciation is well known from numerous fossil finds, not only from archaeological sites but also from such sources as bear caves, so-called mammoth cemeteries (see chapter 9), and all kinds of geological deposits. For all these finds, however, our knowledge of the changing distributions of mammal species large and small is tentative at best. And many animals, especially herbivores, are not particularly sensitive to temperature changes, despite being cold-adapted. Widespread, patchy woodland supported cold-loving animals like reindeer but also warmer-temperature beasts like red deer and wild boar. Human activities complicate the work of the researcher, for people often cut bones into tiny fragments and pursued different species at various times of the year. Small rodents, being far more sensitive to temperature change, are a potentially more reliable source of climatic information. (The color illustrations of Cro-Magnon paintings give an impression of the late Ice Age bestiary.)

Beetles and other insects, when preserved, can also provide important climatic information of a generalized nature; so can earthworms. All of these are esoteric fields of research that rarely receive wide attention.

Our knowledge of the late Ice Age is a mosaic of information from many, often dauntingly complex sources. These, when put together, provide increasingly sophisticated models of a European climatic world that was very different from that of today. Ice cores and stalagmites are likely to flesh out the complex story of the many Dansgaard-Oeschger oscillations that affected human history during the last glaciation.

failed to reach what are now the southern and western shores of the Baltic Sea. This was when extensive open steppe-tundra replaced the woodland and birch forests that had covered much of Europe north of the Alps and the Pyrenees until then. A Neanderthal band on the edge of the steppe would have gazed out over a seemingly desolate landscape of brown scrub and undulating terrain, where bitterly cold winds sliced through the warmest furs and kicked up enormous clouds of dust. The endless plain stretched to the far horizon, relieved only by the occasional shallow river valley and water hole. Few Neanderthals ever ventured onto the steppe in times of extreme cold, except, perhaps, on a quest for bison, which grazed on the tough vegetation that clung to the windswept ground.

From the first cooling, through the serious advance of ice sheets around seventy thousand years ago, to the final deglaciation, European ice sheets, especially those in Scandinavia, were only at their maximum extent for a third of the time, notably during the Last Glacial Maximum, between about 21,500 and 18,000 years ago. Most of the time, the climate experienced by both Neanderthals and Cro-Magnons was considerably milder. Winters might have been long and severe, subzero temperatures commonplace, but most of the time both premoderns and moderns could survive relatively comfortably, thanks, in part, to the continent's environmental diversity. Europe's varied topography provided numerous places of refuge for animals and humans alike, especially during periods of extreme cold.

The most extreme cold of some seventy thousand years ago lasted only about six thousand years, before a sudden warming initiated a sequence of long, fairly mild intervals, apparently triggered by D/O oscillations, occasionally interrupted by brief colder events. There were at least fifteen to twenty short-term events when temperatures were up to 44.5 degrees Fahrenheit (7 degrees Celsius) warmer than during the intervening colder intervals. Sometimes these warm periods were only 3.6 degrees Fahrenheit (2 degrees Celsius) cooler than the average European temperature since the Ice Age. Warm, moist events recorded at Les Echets, in France's Massif Central, brought annual mean temperatures of around 44.5 degrees Fahrenheit (7 degrees Celsius) for several thousand years. (Today's mean is 51.8 degrees Fahrenheit, or 11 degrees Celsius). At Grande Pile, in the Vosges region of northeastern France, mean July temperatures during two warm intervals forty-three thousand and thirty-seven thousand years ago were as high as 61 to 64 degrees Fahrenheit (16 to 18 degrees Celsius). At times, conditions were fairly close to those of today. This period of major climatic instability was when modern humans first came in contact with, and eventually replaced, the Neanderthals.

A final, relatively warm event, shorter and cooler than its predecessor, ended thirty-seven thousand years ago. After that, conditions were much colder, leading to the Last Glacial Maximum. I describe these later shifts in chapter 8. The Neanderthals thrived in a homeland that offered dramatic environmental contrasts. In Mediterranean zones, they enjoyed relatively temperate conditions, especially in sheltered valleys, which were important refuges during periods of extreme cold. North of the Alps and the Carpathian Mountains, they hunted in remarkably diverse and sometimes extreme environments, where forests expanded during warmer millennia and open steppe-tundra extended to the distant horizon on the North European Plain as temperatures fell. In some places, the landscape was arid, treeless, and windswept; in others, it was marked by dark forests. Deep river valleys with fast-moving streams offered shelter to both animals and humans, as well as unusually diverse game and plant foods, a powerful attraction to people who tended to move over relatively short distances.

Their best defense against falling temperatures and deep snow was movement, just as it was for many species of herbivores. The Neanderthals adapted to ever-changing climatic conditions, this time to temperature changes, just as their ancestors had adapted to shifts in aridity in the depths of the Sahara and elsewhere. Such adjustments were second nature to them. Cold times were not necessarily hard times, for lower temperatures meant more open landscape, and the pursuit of migrating herd animals there was somewhat easier than a solitary quest after boar or red deer in a dense oak forest.

There was no typical Neanderthal habitat; environments varied dramatically within a few hundred miles. For example, on the northern plains, mammoths, wild horses, and bison would have been commonplace, whereas the ibex, a wild mountain goat, would have flourished at higher elevations, and the wild ox and boar would have lived in forested valleys, presenting a very different challenge to the hunter. However, even in warmer times, Europe's climate was intensely seasonal, with only relatively short growing seasons of a few weeks or months when plant foods would have been available in significant quantities. Most of the Neanderthal diet had to come from the magnificent bestiary of which they were a small part.

How DID THEY feed themselves? The food resources available to the Neanderthals varied considerably throughout the late Ice Age. During warmer periods glacial summers were longer, the sun stood higher, and the growing season was longer than in the Arctic today. Colder intervals brought much harsher conditions, causing both animals and humans to retreat southward into sheltered valleys and warmer landscapes. None of the cold environments were comparable to those of modern-day northern Canada, Europe, or Siberia. In a world of irregular climatic shifts, many mammal species hunted by the Neanderthals were quite temperature tolerant; they were more strongly influenced by the differences between maritime and dry, often intensely cold continental climates. As far as we can tell from site distributions, the Neanderthals tended to settle in places that offered shelter from strong winds in winter and a diversity of game and plant foods within a relatively limited area, even when migratory animals like reindeer moved away for months on end.

Almost certainly, the Neanderthals were predominantly meat eaters. We know this from the stable isotopes in the bone collagen in Neanderthal teeth. The collagen in a large, worn Neanderthal tooth of fifty-five thousand to forty thousand years ago from the Jonzac rock shelter, in southwestern France, reflects a meat-based diet.[7] The particular group of Neanderthals living there ate mainly bison, aurochs, and wild horses. One uses the word "meat" in the loosest sense, for the people were consuming every part of their prey, not just prime meat cuts. Additional bone-chemistry studies from other sites, such as Saint-Césaire I, in central France, also hint that the Neanderthals relied heavily on flesh, mainly from larger animals. Exactly why, we don't know, but it may be because their weaponry was better adapted to bigger prey than to small animals that were fleet of foot. As always, Neanderthals were opportunistic hunters. Hunters visiting Gibraltar preyed on seals and stranded dolphins.

Seventy thousand years ago, Europe supported a rich and varied mammalian fauna, animals of all sizes; some were adapted to colder conditions and were mainly found in more northern areas, while other, more warm-loving species were confined for the most part to the Mediterranean region. A transitional zone in southern Europe, in sheltered areas like the deep river valleys of southwestern France, supported both cold-loving species and animals that thrived under more temperate conditions. Cave bears, lions, spotted hyenas, and humans flourished in both colder and warmer areas. Wherever Neanderthals settled, there were both large and small herbivores for the taking, as well as familiar predators with whom the human hunters competed.

The fauna varied considerably as the climate shifted. During colder millennia, cold-adapted animals such as the mammoth, the woolly rhinoceros, the musk ox, and the reindeer followed the expanding steppe out of Eurasia and spread across Europe as far as northern Spain. As we have seen, changes from cold to warmer conditions could be very abrupt. When temperatures warmed, tree cover expanded and more heat-tolerant animals appeared to the north. Many of these animals were herbivores, which meant that they followed the changing vegetation to which they were adapted. They would spread slowly into a new area as soon as it was capable of supporting them. As climatic conditions grew colder, the habitat would contract and the animals would retreat, some of them becoming isolated in small pockets before they died out. The process of expansion and contraction never ceased, imperceptible as it was to a human observer on an annual basis. The hunters moved with the game and chose new targets as their prey migrated. Ever-changing ecological realities defined the Neanderthal world.

At one end of the herbivore spectrum were elephants, giant stags, mammoths, and rhinoceroses—the so-called megafauna. These animals had slow reproduction rates and were hard to kill with just spears. The woolly mammoth, *Mammuthus primigenius*, was the giant of the colder bestiary, not a gigantic beast, but standing up to 11 feet (3.4 meters) tall at the shoulders, compared with the 13 feet (4 meters) of an African elephant.[8] Mammoths had short, squat limbs and longer bodies, perhaps an adaptation to plains grazing habitats. Their feet were broader than elephants in temperate environments with laterally extending toes, capable of supporting their heavy bodies in marshy terrain. Even the most expert hunter would probably have had trouble miring a mammoth. They were a hazardous prey at the best of times, with thick hair covering every part of their body except the soles of their feet, even their ears and trunk. So was the woolly rhinoceros, with hair so long that it often trailed on the ground.

The most common animals and most frequent human prey were aurochs, bison, horses, red deer, and reindeer.[9] These were the staples, not only for humans but also for carnivores like hyenas, wolves, and lions. All reproduced more frequently than larger animals, congregated in herds at certain times of the year, and could be hunted singly or by driving entire groups in carefully organized hunts. A single game drive of smaller animals could provide as much meat as one elephant or mammoth at less risk. Larger animals were much easier to kill when rendered helpless by being driven into a swamp or lured into a deep pit dug into a game path. Hunting lesser-sized species still involved substantial risk. The aurochs was a nimble, sometimes fierce animal with long horns and an uncertain temper. European bison could reach massive sizes and were heavily built, with large heads, huge horns, and coats as much as thirty inches (seventy-six centimeters) thick to protect them against arctic winds. Reindeer, like mammoths, were migrants, moving to and from the tundra each spring and fall. Fast runners capable of speeds of up to forty miles (sixty-four kilometers) per hour, Saiga antelope, *Saiga tatarica*, were plains animals that migrated seasonally in enormous herds and had large hooves capable of digging under snow. There were numerous smaller animals, too, among them ibex and roe deer.

Living off the late Ice Age bestiary required not only hunting expertise but also an ability to cooperate and communicate with others. Few Neanderthal hunts were solitary efforts, for no one could tackle a large animal like a bison by themselves. Rather, the hunters worked together, perhaps often in pairs, communicating effortlessly with quiet signals and simple gestures that were as old as hunting itself. Some experts believe women also took part in the hunt. The hunters learned from the school of hard knocks, by example and instinct, by working with more experienced people, using lore that passed from one generation to the next as it had always done. Over the years, each band moved over relatively small territories where both men and women knew every rock and tree and the habits of the animals that lived alongside them. They were constantly on the move, but with one difference from their ancient predecessors. Much of the year still passed in the open, in camps of simple—and long vanished—brush or hide shelters, occupied for a few days, perhaps weeks. Hunting bands had always returned to the same locations again and again: a sheltered river valley, a water hole favored by wild horses, or a place where grasses ripened in early summer. But some Neanderthals now practiced a variation on this routine. They returned to caves and rock shelters in sheltered river valleys again and again, accumulating dense layers of chipped stone, broken bones, and hearths. Though no one knows for sure why they began using them, such places were good refuges against bitter cold and marauding predators, even if huge cave bears sometimes lurked in their depths, and they are a rich bonus for archaeologists studying the dynamics of Neanderthal life fifty millennia ago.

For tens of thousands of years, the even tenor of Neanderthal existence unfolded in seemingly changeless rhythms governed by the movements of game, the seasons of plant foods, and the rotating seasons. There were moments of sudden hunting violence, without question occasional quarrels flared up, ending in fisticuffs or perhaps even death. For the most part, however, each generation passed seamlessly into the next in age-old routines that had deep roots in the remote past.

CHAPTER 4

The Quiet People

WESTERN EUROPE, EARLY SUMMER, seventy thousand years ago. The bison graze peacefully in a forest clearing, knee-deep in the lush grass of the water meadow. Their tails swish back and forth in an endless battle against swarming flies. An elderly male feeds alone, so close to the dark shadow of the trees that the black and brown of its body almost merges with the gloom. The great beast is watchful, suspicious, and alert for predators lurking in the undergrowth.

Two young Neanderthals watch the solitary bison from close downwind, hugging the ground among the trees. They carry stout wooden spears with stone points and are naked, so they can move quickly and in stealth. The hunters slipped into place at first light. Like their prey, carefully selected the night before, they are vigilant, their eyes never still, on the watch for lions looking for an easy kill. Both men seem to melt invisibly into their surroundings, their bodies smeared with mud and grass. Imperceptibly, they creep forward in absolute silence, freezing motionless whenever the bison looks up. Gradually they separate, signaling one another with their eyes, a partnership honed during many hunts. An hour later, the sun is high overhead. The hunters are now so close to their prey that they can almost touch it and would be trampled underfoot if it stampeded. Still they wait, sensing for a moment when the bison is momentarily off guard, its head down.

A quick glance and the man to the left springs to his feet. He leaps atop the beast, plunges his spear deep between its shoulders with a lightning thrust, and jumps clear, using the rearing bison as a springboard. His companion attacks from the other side, thrusting his spear into the animal's hindquarters. The prey roars and rears, stamping blindly on the first hunter, breaking his leg, and then goring him with its horns. He writhes in agony on the ground as the bison moves away, then totters and falls. The second hunter watches closely, keeping well clear of its flailing limbs, then, as it weakens, moves in for the kill.

The rest of the band soon arrive, but the first hunter has already died of his wounds. They swarm over the kill, deftly skinning it, then quickly dismembering the carcass and cutting the meat into strips to dry in the wind. A circle of hyenas watch from a distance, ready to move in when the butchers depart.

FORTY-FIVE THOUSAND years ago, perhaps fifteen thousand to twenty thousand Neanderthals dwelled in Europe and Eurasia, most of them in small groups accustomed to lives of almost-complete isolation. There were fleeting contacts with others, times when men or women would move to another band, or when a hunting accident necessitated cooperation with known neighbors. For the most part, however, these sensitive and talented people lived in tiny worlds, where the unvarying rhythm of daily existence unfolded with the passage of the seasons and the quickening and slowing tempo of climatic shifts. The quiet people lived peacefully among their prey, in a premodern universe that was remarkable for its unchanging verities.

The quiet people: there is no better way to describe the Neanderthals. They went softly about their business, cautiously, watchfully. For all their bulky strength, they moved almost without a sound, mere shadows among the trees and scrub of the late Ice Age landscape. Everyone trod carefully—the experienced hunter, women gathering nuts, boys and girls. They could do nothing else in a world where dangerous predators lurked in the shadows and one could come across a fierce and unpredictable wild ox browsing in the darkness of a forest. Their lives depended on their ability to scout out their surroundings. These were people with acute hearing and what must have been an incredible sensitivity to the movements and sounds of the natural world around them,

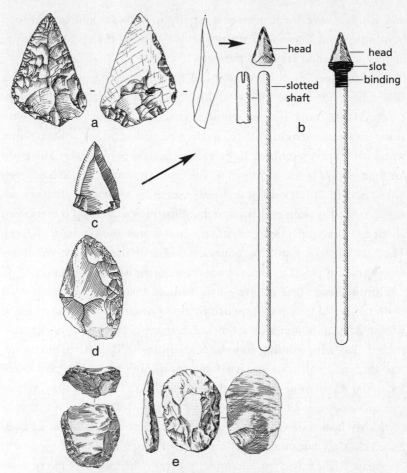

Figure 4.1 *Neanderthal technology and hunting weapons. (a) Points and methods of mounting them to wooden shafts. (b) Scraper (French: racloir), probably used for scraping skins and other purposes. (c) Levallois style "tortoise shell" core (left) and the resulting large, flattish flake. This distinctive technology is named after a Paris suburb.*

the rustle of spring grasses, the light scraping of a deer's foot pawing through thin snow, the gentlest of breezes bringing the distinctive scent of an invisible bison. The Neanderthals could walk, and sense, their way through a late Ice Age landscape in ways that were soon to be compromised by fully articulate speech. The only weapons for both men

and women were their awareness, strength, stalking ability, and long
hunting spears. These were quiet people, who fed themselves by being
inconspicuous and endlessly patient.

This patience must have extended to life in camp, where children
would have played and men would have crafted wooden spears and retouched stone tools. Women would have scraped pegged-out skins with
stone scrapers, then dried them in the sun and wind, or collected firewood from nearby, for hearths must have been kept alight day and night
for fear of predators. For much of the time, the camp would have been
quiet, people sitting calmly, endlessly watching their surroundings, always, at least subconsciously, searching for prey, acquiring information
about the landscape, inventorying the herbivores and predators nearby.
They may have sat still for hours on end, but their eyes were always
moving, always alert in a world where opportunism meant survival.

Hunting with someone armed with an old-fashioned weapon used at
close range is a lesson in patience that you never forget. In my case, it
was an African hunter with a flintlock musket once used by a Victorian
soldier. We went hunting impala, a medium-sized, fast-running antelope that thrives in grass and is famous for its high leaps. The hunt lasted
for hours as we moved quietly through the bush, slipping from tree to
tree, occasionally pausing to examine impala droppings on a narrow
track. My host stalked like a wraith, making never a sound, while I followed clumsily but under the illusion that I was silent. We lost a shot at
one beast when a twig cracked under my foot when we were thirty yards
(twenty-seven meters) away. The next time, we got within ten feet (three
meters) of a solitary male, freezing into immobility each time the impala looked up. It seemed to take minutes for the hunter to aim, but in
fact it was a few seconds. He fired and missed. The impala ran off, and
we had to start all over again.

It was then that I realized what hunting with a wooden or stone-
tipped spear meant: not just a quick thrust from a relatively short distance, but a fatal jab with a razor-sharp weapon delivered at point-blank
range. A successful Neanderthal hunt depended on the stalk, a careful
approach from downwind that took full advantage of an intimate
knowledge of the quarry's habits and reactions. Above all, the hunters

had to be quiet and unobtrusive. They worked as teams, so familiar with one another that they could communicate with gestures. Success came from deep wells of patience and endless waiting, from an acceptance of regular failure. The hunters knew full well that to wound their prey was not enough. They then had to track the bleeding animal and either kill it or recover the carcass before lions or other predators beat them to it. Like modern-day hunter-gatherers, they must have been stoic in the face of accidents and hunger, measuring risks carefully and taking advantage of every opportunity to obtain food.

How much do we know about Neanderthal hunting abilities? What little we know comes from artifacts and animal bones, as well as intelligent theorizing. There seems little doubt that men, women, and even children were adept at hunting even quite large prey. They had to be, for their bodies demanded thousands of calories a day in a world of cold temperatures. During the cold millennia after 75,000 years ago, they appear to have ambushed migratory animals like reindeer, presumably in spring and fall, a form of hunting that would have required advance planning. Or did it? The American archaeologist Lewis Binford has described Neanderthals as foragers, people who roamed opportunistically over the landscape exploiting any resources they might encounter, something quite distinct from the kinds of careful forward planning and anticipation characteristic of modern humans.[1] Their instinctual knowledge of the environment would have surely led them to frequent places where, for example, migrating reindeer herds crossed river fords, circumstances where advance planning and elaborate ambushes were unnecessary. Some of their meat may have come from scavenging predator kills, but how much would have been left after a pride of lions and then hyenas had feasted is questionable. Nor would they have felt comfortable chasing off feeding predators from a kill, a high-risk pastime at best. The hunters would have sometimes come upon animals that had just died of natural causes, or beasts that were tottering at life's end and were easy to approach and kill with a quick blow or a jab with a wooden spear. Some Neanderthal caves in western Italy have yielded parts of the skulls of older animals, as if the people were preying on such beasts during the last interglacial, some 120,000 years ago.

But the unobtrusive, quiet Neanderthals thrived not because they were scavengers and opportunists, which they were, but because they were consummate hunters with an ancient weapon: the wooden spear.

Neanderthal hunting weapons were little more elaborate than those used at Schöningen more than three hundred thousand years earlier, except for the use of carefully shaped stone points attached to wooden spear shafts (Fig 4.1a). The weapon of choice would have been a stone-tipped or wooden thrusting spear—the latter with a fire-hardened point—the range and penetrating power of which would have been limited by the strength of the hunter's arm. Twenty-five feet (eight meters) or so would have been the absolute outside range.[2] To kill a larger beast such as an aurochs or bison would have required a much closer approach, achieved either by stalking or by driving the prey into a swamp or narrow gully, a task involving the cooperation of an entire band or even more people. Wooden throwing sticks like the example from Schöningen, the ancient equivalent of the Australian boomerang, would have been effective against smaller prey. Every hunter must also have carried a wooden club, the most basic of weapons. Short of harvesting reindeer migrations and opportunism, some hunters may also have been fleet enough of foot to run down smaller deer and other animals, tackling them to the ground and then dispatching them with a club. Every form of hunting, even scavenging, was high risk. To kill bears, lions, or wolves for their hides must have required extreme courage and skill. It would have been suicide to attempt to spear them when they were attacking, except in desperate self-defense. The hunters probably killed them while they were hibernating or sleeping in their dens.

We have surprisingly little scientific data on their hunting. Neanderthal bands visited the Combe Grenal cave, in France's Dordogne, between about 115,000 and 50,000 years ago. No less than fifty-five closely packed occupation layers document repeated visits, and each visitation was carefully excavated by the great French archaeologist François Bordes during the 1950s and 1960s. He recovered thousands of stone artifacts and animal bones but died before he could publish his findings on his excavations. Philip Chase of the University of Pennsylvania Museum studied the

Combe Grenal animal remains.[3] Herbivores dominate the occupation levels, including horses, red deer, and reindeer. By studying the cut marks on the bones, Chase was able to show that red deer and reindeer were hunted deliberately. Red deer were a favored prey during warmer conditions between about 115,000 and 100,000 years ago, while reindeer replaced them thereafter, when colder conditions replaced forest with steppe except in sheltered river valleys. Only small numbers of these animals appear in each layer, as if they were stalked and killed individually, with most of the carcass carried back to the cave for butchering. In contrast, among the other herbivore bones, such as those of horses, are relatively large numbers of skulls and jaws, which may mean that these animals were scavenged from carnivore kills. Perhaps the predators that killed them were unable to crack open the skulls to extract the highly nutritious brain and tongue, so the heads were sometimes the only parts of the bodies left behind.

At Combe Grenal, the hunters sought out animals on a one-to-one basis, but the constant environmental changes meant that they took whatever they could find within range of the cave. This would have required at least some advance planning, especially the sharing of intelligence about nearby game sightings.

Elsewhere, the Neanderthals certainly engaged in mass killing, an activity that involved deliberate planning and timing and close cooperation among all members of the band, and perhaps several other families and their kin. One hundred and twenty-five thousand years ago, a Neanderthal group living in a cave at La Cotte de Saint Brelade on Jersey in the Channel Islands, drove both mammoths and rhinoceroses over a cliff, then killed and butchered the helpless beasts at the foot, dragging selected body parts into the cave away from marauding carnivores.[4] Excavators found stacks of bones still lying at the foot of the cliff face. Likewise, far to the south, a massive concentration of bison bones lies close to a steep riverside cliff at Mauran, in the French Pyrenees.[5] Fifty thousand years ago, Neanderthal hunters drove their prey off the escarpment to their deaths, a task that must have involved significant numbers of hunters working closely together.

Like their predecessors, the Neanderthals trod lightly and relied as much on their powers of observation and quiet imprint on the land as they did on their simple technology.

CLICK, CLICK—THE distinctive sound of hammerstone against flint would have echoed through the still morning air of sheltered river valleys. It was a commonplace sound to people who relied on fine-grained stone like flint or chert and wood for tools and weapons. The entire Neanderthal tool kit probably numbered no more than a few dozen tools and weapons, nearly all of which were fashioned with stone tools or were stone themselves. Unfortunately, the brutal realities of soil chemistry mean that only the most durable Neanderthal artifacts have survived. When we are lucky, so have the bones of the animals they hunted. We know of their lives, of their ingenuity, almost entirely from stones. When I was a student, a multitude of stone tools and a handful of fossils defined the Neanderthals, the former classified into ever-proliferating, meaningless tool types that changed from one researcher to the next. Today's scholars draw on a far broader range of sources, but much of what we know about Neanderthal history is still from stone tools, the most durable and often least informative of all artifacts.[6]

Neanderthal technology owed almost everything to toolmaking traditions that had developed in much earlier times. The earliest human tools of two million years ago were little more than sharp-edged flakes and crude, jagged-edged choppers, though making them required a sense of the properties of different rocks and at least some rudimentary knowledge of the best ways to hold a rock when flaking it. Innovations were few and far between in a human world where wooden spears, simple digging sticks, stone scrapers, and multipurpose stone axes sufficed for foraging, hunting, and butchering for hundreds of thousands of years (figure 4.1b). Then, some 250,000 years ago, a new stone tool technology developed in Europe and elsewhere, which involved careful preparation of stone nodules, or cores, before a single flake was struck off them. This simple innovation, the prepared stone core, was a cornerstone of Neanderthal tool kits . . .

The young man sits on a hide, enjoying the warm summer sun. He picks up a nodule of flint collected the day before from a nearby cliff. After turning it carefully in his hands, he picks up a round hammerstone and knocks off a large flake. The white cortex of the stone falls away, exposing the black flint inside. Deftly, the stoneworker uses the exposed surface as a platform to remove a series of flakes from the block. The lump fractures without warning from a hidden fissure. He throws it to one side and picks up another one. Again, he subjects the rock to an appraising gaze, then strikes off a large flake. This time, he fashions the rock successfully into a profile like an inverted tortoiseshell: nearly flat on top, gently curved on the lower surface (figure 4.1c). Now he gently hammers off a few small flakes at one end to make a keying surface for what is to come. He massages the rock in his hands, appraising the shape, feeling the quality of the stone. Then he places it on one end between his feet on a piece of bison bone. A single, sharp blow, and a large oval flake falls away from the surface of the core. Nearby lies the wooden spear shaft with a snapped flint point, which he will replace with a new one carefully shaped from the oval flake . . .

Prepared cores, a technology whereby a stone nodule was carefully shaped before being turned into tool blanks, had the advantage of producing quite thin flakes of more or less standardized shape and dimensions, which could be used for all kinds of purposes with minimal additional trimming. For instance, a large oval-shaped flake like the one made by our stoneworker could be used for cutting and scraping, while other cores of different shapes produced smaller, more triangular forms that could be mounted on wooden spear shafts. Most flakes produced from prepared cores also had virtually continuous working edges all the way around. Neanderthal tool kits became even more versatile when stoneworkers began using another form of prepared, disklike core that allowed the toolmaker to produce not just one flake but several, all roughly the same size and shape. These were the earliest relatively standardized artifact blanks of which we know. Following established practice, nineteenth-century archaeologists called this distinctive technology the Mousterian, named after Le Moustier cave, in the Vézère Valley, where it was first identified (see the box "Aurignacian, Gravettian, Solutrean, Magdalenian . . ." in chapter 6 for terminologies used in this book).

Deciphering stone tool technology

Stone technology is as old as humanity itself—some 2.5 million years. In the final analysis, toolmaking is a linear process. The stoneworker acquires the raw material, prepares a lump of stone (the core), and then reduces it by striking off flakes. The earliest artifacts were sharp-edged flakes used for a variety of purposes, struck off their parent core with a few blows from a suitable hammerstone. Even these simple artifacts required the right kind of rock, the ability to visualize a tool in three dimensions, and a "feel" for the properties of the stone.

Both Neanderthals and Cro-Magnons chose fine-grained stones like flint or other hard, homogeneous rocks for their artifacts. All these rocks break in a predictable way, like glass, which enables you to knock off flakes. At first, premoderns used stone hammers. Then they turned to antler, bone, or wood to trim the edges of their hand axes (see figure 2.3). The softer hammers produced thinner flakes. By one hundred thousand years ago, Neanderthal stoneworkers had gone a stage further. Now they preshaped their cores to produce either one or several flakes. To do this successfully required careful preparation of the striking point, often called the striking platform, to ensure an accurate blow when the flake was struck off. The result: thinner flakes of more standardized triangular and other shapes, which could be mounted as spear points and used for other composite tools. Cro-Magnons refined tool technology even more with punch-struck blades struck off conical cores, a technology described in chapter 8. They relied on a far wider range of specialized tools.

How do we decipher all this changing technology, its end products the most durable of all archaeological finds? The stone tool "types" referred to in this book are based on what the classifier considered to have been their original use. When we talk about scrapers with steep, angled edges, points with sharp edges, and borers, such classifications suffer from several disadvantages. First, these classifications rely on what we think they were used for, although our instincts are probably correct. Second, the nature of stone technology is such that it's very hard to make truly standardized, mass-produced tools, so variation in, say, scraper forms is very wide.

The best artifact classifications have come from scholars who have acquired firsthand experience in making all kinds of stone artifacts. Louis Leakey, the famed paleoanthropologist, was an expert stone-tool maker who skinned antelope with stone flakes. François Bordes could turn out a finely crafted hand ax in two minutes or less. Today, lithic technologists abound. At this point, we collectively know as much about stone-tool making as our late Ice Age forebears, if not more.

Today's lithic analysis now goes far further than the mere study of tool forms. Watch someone making stone tools and you'll find that the person is sitting in the midst of a pile of ever-accumulating debris: chips, flakes, abandoned nodules, and discarded hammerstones. These by-products of toolmaking are as informative as the artifacts themselves, especially if you attempt to refit the debitage (waste material) back to its native core. Refitters are scholars of unlimited patience—completing one artifact can take months—but they can sometimes achieve extraordinary results, establishing from the direction of hammer blows that a knapper (stone worker) was left-handed, piecing together hundreds of flakes large and small from a single core scattered over several square feet of a campsite, or ascertaining that a stoneworker made tools at one location, then carried a core to a nearby hearth and fashioned a blade for a quite different purpose.

Where did the rock come from? Sourcing rocks to their native outcrops yields valuable information on trade and mobility. Petrological and spectrographic analyses use microscopes and thin sections to identify the distinctive characteristics of rocks from different locations, which are, of course, matched in the artifacts made from them, some of them over fifty miles (eighty kilometers) away.

Many of the artifacts and flakes in both Neanderthal and Cro-Magnon sites were broken during their short, arduous lives. Both peoples used nearly every tool they made for a variety of purposes: for killing animals, for skinning them, perhaps for cutting leather or working timber. Many of these activities left telltale use-wear patterns on working edges that can be identified with a microscope, then pinned down to specific activities like stripping meat from bones or cutting, by comparing them with edge wear resulting from controlled experiments with a variety of activities. Use-wear analysis is not a matter of just studying individual artifacts, although obviously that tells us what they were used for; the most significant results come from much larger samples, where statistical treatment factors out chance utilization. Few such studies have been completed. None of this research is headline grabbing, but it is nice to know that some Neanderthals were left-handed.

All stone-tool making depended on reliable sources of good-quality volcanically derived rock, the finer grained the better. For people like the Neanderthals, who had an encyclopedic knowledge of their territories, locating good stone outcrops was a matter of routine. What is striking, however, is that they apparently made only rare efforts to obtain tool-making stone from afar and tended to rely on local supplies. When they lucked upon good-quality flint and other such rocks, they used the tools they made from them again and again, constantly resharpening and modifying them until the artifacts were essentially useless for any purpose and were thrown away.

The Neanderthals employed stone tools for cutting and scraping and, sometimes, for weapons, but the edge polish on their cutting surfaces reveals that many of them were used for woodworking. Hunters may have occasionally fashioned a sharp stone point for a spear, but they appear to have used most of their stone tools to fashion wooden artifacts of all kinds. We know from the Schöningen weapons of four hundred thousand years ago that spear makers toughened their spears by charring the points and shaped the shafts with the weight near the tip to give them weight, range, and a lethal spinning motion in the air. No question, the Neanderthals did the same, for the spear was, once again, the only weaponry at hand.

Neanderthals depended heavily on stone and wood, but not on bone and antler. They relied on flaking and polishing, on scraping and cutting, but lacked the delicate chisels that enabled their modern successors to unlock the technological potential of the antlers and bones of the animals they hunted. A few sites have yielded fragments of bone or antler used to strike flakes from stone tools and a few crude bone implements, but there is none of the elaboration of the delicate artifacts manufactured by the Cro-Magnons. The stone and wood technology used by the Neanderthals was both economical and efficient, and honed to a fine quality, but it was limited by the cognitive abilities of its users.

THE QUIET PEOPLE: the Neanderthals were surprisingly human in their daily life, yet mired in a premodern routine that had changed but little since a few humans had hunted across Europe hundreds of thousands of years before. They preyed on the bestiary of which they were a part, were subject to the same ecological pressures as their quarry, and sometimes fell victim to their fellow predators. And theirs was not an idyllic life. Even close-knit bands experienced periods of resentment and tension when tempers flared and, inevitably, people came to blows— perhaps a quarrel over leadership or women. There are telling signs of interpersonal violence on some Neanderthal bones. Even fleeting contacts with neighboring groups could turn violent, over territory or, again, women. Spears might fly, clubs might be wielded, leaving wounded or dead men on the ground. Such behavior must have been instinctual, like that of competing animals defending their young or their boundaries, with the difference that powerful emotions may have sometimes played a role.

Some Neanderthals also consumed human flesh. Between 120,000 and 100,000 years ago, a group visited the Moula-Guercy cave, located twenty-six hundred feet (eight hundred meters) above the Rhône River in southeastern France's Ardèche, just as the last glaciation began.[7] Three hearths and a crude stone wall chronicle the sporadic occupation, as do stone scrapers and other artifacts and the bones of red deer and wild goats. Seventy-eight human bones from at least six individuals, including two young children, were also found at the site. The comminuted pieces lay through some nine feet (three meters) of the deposits, almost all of them displaying exactly the same kinds of butchery marks as those on the red deer bones—multiple cut marks, fractures resulting from hammering to extract marrow and brains, crushing of spongy bone, and peeling. The victims were defleshed and disarticulated before their bones were placed on an anvil and broken open with sharp blows. These particular Neanderthals were cannibals, but they were not unique. At El Sidrón cave, in northern Spain, the remains of at least nine Neanderthals were found in a limestone chamber. A short time after they died some forty-three thousand years ago, their bodies tumbled far underground as the ground collapsed beneath them. The bones of five young adults, two adolescents, a child of about eight, and a toddler lay in a jumble. All had suffered from malnutrition, detected from layering in their teeth. They had been butchered for their brains and marrow.

Neanderthals have been accused of cannibalism for nearly a century on the basis of fractured bones at other sites, including Krapina, in Moravia, in the Czech Republic, where, however, the people cleaned human bones, perhaps before burying them. They did not extract the marrow. Why cannibalism? Did Neanderthals routinely, if rarely, consume human flesh? Or were these occasional episodes, part of simple death rituals, in which, perhaps, the living acquired strength from the newly dead? We will never know. We do know that the Neanderthals were the first humans ever to bury their dead, though whether it was because of a practical need to protect the bodies from predators is unknown.[8] The burials themselves tell us little. Most Neanderthal graves have been found in abandoned occupation levels in caves and rock shelters. Some of the graves have contained stone tools or animal bone fragments, but they could easily have been buried accidentally with the corpse as the pit was filled in. Despite claims to the contrary, which need not detain us here, there is nothing to suggest that Neanderthal burial was much more than a convenient way of disposing of the departed, an essential defense strategy, especially in winter, for people living in caves often frequented by carnivores.

One reason for the apparent lack of ritual may be that Neanderthals lacked completely articulate speech and the constant give-and-take of the conversation that goes with it. However, they were certainly highly effective communicators on a practical level. They had to be, for success in the hunt depended on efficient communication between stalkers facing acute potential danger.

THE NEANDERTHALS HAD brains as large as those of modern humans, but behaved in quite a different fashion, foraging opportunistically across the landscape in such a way that we must assume that they lived largely without language. They used their brains in a distinctive way. Neanderthals had a sophisticated communication system that had evolved over many thousands of years of selective pressures, especially for cooperation with others, not necessarily a single person, but several others. Habitual cooperation was vital in a world of severe, often rapidly changing climatic conditions, especially when all kinds of foraging were opportunistic and when food was shared between all members of a group, whether their individual hunts had been successful or not.

Promoting cooperation and enhancing it must have been a fundamental part of Neanderthal life. The archaeologist Steven Mithen has developed a provocative theory about Neanderthal communication. He believes that making music together, whether for dancing or a simple hum or chant, was a way for Neanderthals to promote future interaction and cooperation in future hunts.[9] He speculates that a propensity for making and listening to music has been part of the human genome for a very long time indeed. Language transmits information; music expresses and induces emotion. For this reason he proposes that some form of musical expression may have been a foundation of Neanderthal existence.

Mithen calls the Neanderthal communication system "Hmmmmm"— "holistic, manipulative, multi-modal, musical, and mimetic in character." He believes that rudimentary forms of this system had developed much earlier, but that the Neanderthals developed it to an advanced and re-markably successful level, so much so that people survived for a quarter of a million years, through periods of dramatic environmental change. In short, says Mithen, they were "singing Neanderthals." Their songs lacked words, but they were intensely emotional. There were "happy Nean-derthals, envious Neanderthals, guilty Neanderthals, grief-stricken Nean-derthals, and Neanderthals in love." Mithen speculates that they needed such emotions because their way of life required not only constant and in-telligent decision making but also extensive social cooperation. "Hmm-mmm" is, of course, nothing more than intelligent speculation, but it certainly makes logical sense.

The Neanderthals lived in tiny bands, just like their ancestors had done, sparsely distributed over a vast and varied landscape. They rarely came into contact with strangers and must have known each other intimately—through family histories, social relationships, and the nature of day-to-day activities. There were none of the large gatherings, long-distance trade relationships, and specialized activities, such as sewing tai-lored garments, found among their successors. With such small groups, it's hardly surprising to find each short-lived generation making exactly the same stone tools as their predecessors.

Did the Neanderthals have the ability to pass complex information from one generation to the next? Like music, language is an intangible, which can only be studied indirectly, notably by looking at tiny bones like the hyoid, attached to the cartilage at the larynx, which anchors the muscles needed for speech. Only one Neanderthal hyoid bone has been recovered, on a sixty-three-thousand-year-old skeleton from the Kebara cave, in Israel, but it is lower in the throat than that of *Homo sapiens*.[10] The dimensions of the hypoglossal canal, which carries the nerves that run from the brain to the tongue, are equivalent in Neanderthals to those of the hypoglossal canal in modern humans, while the dimensions of the canal in the thoracic vertebrae, which house the nerves for controlling the diaphragm and breathing, are also similar. Thus, we know that Ne-anderthals had the same motor control over their tongues and breath-ing as we do. They also appear to have had the same sound-perception structures as people today, as well as some of the hardware for speech. The bones of two of the Neanderthals at El Sidrón have yielded the FOXP2 gene, which contributes to speech and language ability by acting on both the brain and the nerves that control facial muscles.

For all of these anatomical features, there are reasons to believe that Neanderthals did not possess fluent speech. For one thing, the Neanderthals lived in small, intimate groups. They shared knowledge and experience about their environment and their tiny social world, always with a common purpose. Mithen argues that with "Hmmmmm" they would not have needed any new forms of utterance such as those produced by speech: "They didn't have much to say that had not either been said many times before or could not have been said by changing the intonation, rhythm, pitch, melody, and accompanying gestures of their widely understood, simple utterances."[11]

Mithen believes that we can identify a few places where Neanderthals danced and performed, among them a deep chamber several hundred feet from the entrance of the Bruniquel cave, in southern France, where a sixteen-by-thirteen-foot (five-by-four-meter) quadrilateral structure is demarcated by pieces of stalactite and stalagmite. There, in total darkness, Neanderthals may have danced and sung familiar rhythms by the light of a fire or flickering brands. Sounds would have echoed off the walls and resonated in the dark space, as their shadows ebbed and flowed against the walls. A fragment of burned cave bear bone in the middle of the structure dates to 47,600 years ago. Bruniquel's walls would have been a superb "canvas" for cave paintings or engravings, but there is no art there. A few other sites are notable for the limited areas where Neanderthal visitors camped and left occupation debris, leaving open space nearby, where, perhaps, they danced—but this is, of course, pure speculation.

Generations of archaeologists have failed to discover any intentionally modified Neanderthal objects that could be classified as art or artifacts with symbolic meaning quite unrelated to their form, except for a few scratched bones and stones. They may, however, have painted their bodies using powdered manganese oxide mixed with water or other liquids. Body painting does not, however, necessarily imply symbolic meanings. It may have been used for camouflage on the hunt, or to emphasize personal appearance, perhaps as a form of sexual attraction.

The Neanderthals may have attributed symbolic meanings to animals, trees, prominent landmarks, and other natural features, though, of course, we have no means of knowing if this occurred. It seems unlikely that they did so at anything more than the basic level, for to attribute such meanings to objects of any kind requires language. We do this every day. Even sitting at my computer, I can see at least half a dozen objects with symbolic meaning for me at a glance. As Mithen argues, it seems inconceivable that the Neanderthals flourished for more than two hundred thousand years without using language for this purpose, especially since they lived in very challenging environments and often on the edge of survival.

However, the most powerful argument against language is the long-term stability of Neanderthal culture. The tools they made and their way of life survived virtually unchanged between about 250,000 and 30,000 years ago. Language is a force for change, for exchanging ideas, and for complex thought. Of course, their culture was extremely diverse and their behavior often complex. They had a high appreciation of good toolmaking stone and, presumably, of other raw materials. But there were no innovations, just a narrow repertoire of ancient technologies that sustained them for thousands of years. No Neanderthal ever invented a needle and thread, a harpoon, or a bow and arrow. Such inventions were the domain of language-using modern humans.

What, then, about the Neanderthal mind? In some respects, the Neanderthals appear remarkably modern. They fashioned complex artifacts, hunted all manner of game, and survived through major climatic shifts for more than 200,000 years. Yet they lacked speech, and their culture was effectively unchanged over hundreds of thousands of years. There are no signs that they were capable of symbolic or theoretical cognition. Mithen has another theory.[12] He argues that the Neanderthals had what he calls a "domain-specific intelligence." They had vast stores of knowledge about the natural world and reacted to it in near-modern ways. They had excellent technical skills for making sophisticated tools; their social relationships were both complex and continually maintained. But they were never able to use their technical skills to make artifacts to mediate these social relationships, such as clothing or jewelry. Nor did they design specialized hunting weapons like those of modern humans, because they were incapable of bringing together their technical expertise and knowledge of their prey in a single conceptual thought. They never fabricated a weapon specifically for use against a particular animal, nor did they make the kinds of multicomponent artifacts used by their modern successors.

As Mithen points out, what the Neanderthals lacked were the additional neural circuits in the brain that would have connected toolmaking, socializing, and human interactions with the natural world. The archaeologist Thomas Wynn and the psychologist Frederick Coolidge have suggested that the missing circuits were those related to working memory, those that would have allowed them to retain several kinds of information in active attention simultaneously.[13] This was why, Mithen speculates, the Neanderthals expressed complex emotions and information about the natural world through sophisticated iconic gestures, dance, vocal imitation, and song, but their communication system was one of relatively fixed utterances that helped perpetuate conservative thought and a static culture.

Neanderthal life was full of decisions that were matters of life and death. Such decisions involved every aspect of daily existence: where to hunt and when, what animals to pursue, and potential partners for the chase. Who would mate with whom? Who would care for infants, the seriously injured, or the children of someone killed in a hunting accident? Such decisions involved the need not only to process information but also to feel emotion. Optimal decision-making depends on emotion, which means that the Neanderthals were probably highly emotional people and expressed their emotions through gesture and utterance, as well as body language. Their survival depended on cooperation, on working together during a hunt or while collecting plant foods, on caring for everyone's well-being. In a world where contacts with others were very limited—although obviously there were some that preserved genetic viability and allowed bands to survive as social units—creating and maintaining a social identity was all-important, perhaps achieved through communal singing and dancing.

The slow-paced world of the quiet people changed drastically about forty-five thousand years ago, when Cro-Magnon newcomers shattered the foundations of Neanderthal existence.

CHAPTER 5

The Ten Thousandth Grandmother

"IMAGINE A DINNER TABLE SET for a thousand guests, in which each man is sitting between his own father and his own son. At one end of the table might be a French Nobel laureate in a white tie and tails, and with the Legion of Honor on his breast, and at the other end a Cro-Magnon man dressed in animal skins and with a necklace of cave-bear teeth. Yet each one would be able to converse with his neighbors on his left and right, who would either be his father or his son. So the distance from then to now is not really great."[1] The late Finnish paleontologist Björn Kurtén's hypothetical banquet places us at the same cognitive table as the Cro-Magnons, even if you might, in fact, need two thousand guests. A wide cultural divide separates the Nobel laureate from the skin-clad guest at the other end of the table, but they enjoy the same cognitive abilities. Both are *Homo sapiens*, the "wise person."

We "wise people" possess formidable intellectual powers and reasoning skills. We speak fluently and mellifluously, not only communicating about practical things but also projecting a myriad of feelings, subtle meanings, and emotions. We plan ahead and look at our surroundings as we interact with them, negotiate with one another, cooperate, argue, and fall in love. We fight and hate and, above all, are opportunistic but also innovative; we look at the world in symbolic terms. We are humans with a consciousness and a self-awareness that impact every part of our lives.

In his *Man's Place in Nature*, Thomas Henry Huxley compared Neanderthals and African apes. Huxley's essay appeared in 1863, soon after the sensational discovery of the Neanderthal skull, but before the

Cro-Magnon excavations in 1868, which produced the modern successors to much more ancient humans. For generations, scientists assumed that the Cro-Magnons were the first modern humans anywhere, that the Neanderthals were their ancestors. These scientists lived in an ethnocentric world, which revolved around Europe, the birthplace of modern industrial civilization. For them, the "question of questions" concerned early human ancestry and a putative "missing link" between apes and people.

In his *Descent of Man*, published eight years later, Charles Darwin pointed to Africa as a possible cradle of humanity. A century and a half later, we know for sure that humankind evolved south of the Sahara at least 2.5 million years ago. Many details remain uncertain, but we've answered Huxley's most fundamental questions about our ultimate beginnings. Today, the "question of questions" about human origins surrounds not the earliest humans of all but another mystery of fundamental importance: where, when, and how did *Homo sapiens* first appear? We cannot understand the Cro-Magnons without understanding where their ancestors came from.

AS NOTED IN chapter 1, the search for the origins of modern humans shifted from Europe to the Near East before World War II. In 1929, a Cambridge University archaeologist, Dorothy Garrod, excavated a series of caves in the slopes of Mount Carmel, near the coast of what is now Israel.[2] Garrod was an experienced cave excavator and well-schooled in Cro-Magnon artifacts from European rock shelters. Over a series of field seasons conducted on a shoestring budget, she dug into two caves: Mugharet el-Wad and et-Tabun. As she dug ever deeper into el-Wad, she uncovered occupations left by *Homo sapiens*, then those of even earlier hunters, who used Mousterian technology, fabricated by Neanderthals. Et-Tabun produced an even earlier sequence and a single Neanderthal burial (figure 5.1).

Neanderthals, then modern humans: the two Mount Carmel caves produced no surprises. Then biological anthropologist Ted McCown of the American School of Prehistoric Research, in Cambridge, Massachusetts, excavated an unprepossessing rock shelter known as es-Skhul,

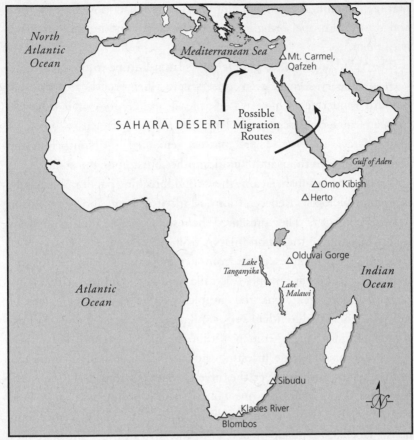

Figure 5.1 *Map showing some locations mentioned in chapters 5 and 6, also some possible routes out of Africa.*

three hundred feet (one hundred meters) east of el-Wad.[3] He uncovered at least eight burials associated with Neanderthal tools. The Skhul skeletons crouched in small graves, their bones a striking contrast to the familiar Neanderthals of Europe. They were slender people, apparently about forty thousand years old, whose bones displayed a mixture of archaic and modern features. McCown and Garrod believed these were the transitional people between Neanderthals and moderns, as if the critical changeover from premodern to *Homo sapiens* had taken hold not in Europe but in the Near East. In 1938, Garrod published a famous paper in which she argued that the ancestry of the Cro-Magnons lay in that region. From there, small numbers of them had moved into central and western Europe, bringing new technologies and hunting methods with them.[4]

The Skhul finds turned people's eyes from Europe to the Near East. Garrod's theory seemed even more secure when French archaeologist Bernard Vandermeersch excavated at least seven *Homo sapiens* burials from a Neanderthal occupation level in the huge Qafzeh cave, inland of Haifa, Israel, in 1965. There matters remained until mitochondrial DNA and our ten thousandth grandmother burst upon the stage.

The genetic bombshell came in a groundbreaking paper published by molecular biologists Rebecca Cann, Mark Stoneking, and Alan Wilson in *Nature* in 1987.[5] They presented the results of more than seven years spent collecting mitochondrial DNA from the placentas of newly born children. Their samples came from 147 individuals, whose ancestors lived in Africa, Asia, Europe, Australia, and New Guinea. After elaborate laboratory treatment, the samples yielded 133 distinct types of mtDNA. Some children had very similar sequences, as if they had descended from a single woman within the past few centuries. Others shared a common female ancestor, who had lived thousands of years ago. In the abstract to their paper, the three geneticists wrote, "All these mitochondrial DNAs stem from one woman, who is postulated to have lived around 200,000 years ago, probably in Africa." Inevitably, science journalists labeled this shadowy ancestor "the African Eve." She was, they said, our ten thousandth grandmother, when, of course, she was in fact part of a small population.

A storm of controversy greeted the *Nature* paper. Biological anthropologists divided into two camps. One school of thought, known as multiregionalists, argued that *Homo sapiens* had evolved from earlier humans in several regions of the Old World. Their opponents supported the geneticists and what became known as the "Out of Africa" hypothesis, the notion that *Homo sapiens* had originated in tropical Africa, then spread from there across the late Ice Age world. A generation later, the furor has subsided in the face of new, even more sophisticated research involving both mtDNA and the Y chromosome. The genetic evidence is overwhelming. *Homo sapiens* evolved in Africa,

Chasing Adam and Eve? Mitochondrial DNA and Y chromosomes

Since the 1980s, molecular biologists have studied modern-human origins, as well as the genetic histories of modern populations. Two approaches have worked well. The first entails considering the genetic variations between living populations, the notion being to identify the most recent common ancestor of everyone living today. The second involves isolating DNA sequences from fossil *Homo sapiens* and premoderns such as Neanderthals.[6]

Molecular researchers create phylogenetic trees based on data obtained from different parts of the genome and on the variability within these systems that one can observe in living societies. In most cases, they use variability in DNA sequences to create the family trees, the earliest common ancestor being the deepest point in time where the branches meet. Finding the primordial ancestor using phylogenetic trees is ultimately an exercise in statistical probability, based on assumptions about such factors as population size and so on. It is one thing to identify the earliest starting point, quite another to assign a date to him or her. This is where the researcher has to determine the rate of genetic change and calibrate a molecular clock.

It should be stressed that phylogenetic trees give no indication of the behavioral or physical changes involved in the establishment of our species. That is where archaeology and skeletal anatomy come in. The genetic trees largely come from studies of mitochondrial DNA variation in living people. Mitochondria are the engines of cells, for they metabolize food and water into energy. They also maintain their distinctive DNA over tens of thousands of years. Mitochondrial DNA has only about sixteen thousand paired subunits of nucleotides (bases), is much easier to analyze than nuclear DNA, and has one priceless advantage. It is inherited only from the mother and passes intact from one generation to the next. This allows scientists to focus on the changes caused by mutations and mutations alone. By measuring the number of mutations that have taken place in the mtDNA of primates, whose evolutionary divergence millions of years ago has been dated from fossil bones, researchers have developed an mtDNA molecular clock.

Mitochondrial DNA mutates much faster than nuclear genetic material, changing every few hundred years. Thus, it can be used as a gauge of short-term evolution and especially as a timepiece for measuring when modern humans diverged from a common ancestor. Rebecca Cann and her colleagues published an evolutionary tree that showed that modern humans, *Homo sapiens*, had originated in Africa between 90,000 and 180,000 years ago.

More-recent analyses of the entire mtDNA genome by molecular biologist Michael Ingman and others have confirmed the findings of the original study. We now know that three of the deepest branches of the mtDNA tree are exclusively African, the next deepest being a mixture of Africans and non-Africans. All non-African DNA branches are of a very similar depth. Ingman and his colleagues believe that the mtDNA lineage evolved for some time in Africa, followed by an out-migration by a small number of people. A population bottleneck resulted, followed by a population expansion. All later European and Asian *Homo sapiens* lineages originated in this small African population. The researchers also refined the chronology, dating the most common recent ancestor to about 171,500 years (plus or minus 50,000 years) ago. The date of the earliest branch that includes both Africans and non-Africans is 52,000 years ago (plus or minus 27,500 years).

Y chromosomes are, in many respects, the male equivalent of mtDNA. The Y chromosome transmits across generations, but in the male. Much of it undergoes recombination, the rearrangement of genes that occurs during meiosis. The portions that do not are used to construct phylogenetic trees. In a landmark paper published in 2000, twenty researchers studied a worldwide sample of Y chromosomes of men from dozens of populations on every continent. Using

> the same methods as were used in earlier mtDNA studies, they constructed a male family tree using the splits in the ancestry of the chromosome. The result was the same as that for mtDNA: the root of the tree lay in Africa. However, the "African Adam" lived not some 150,000 years ago but only 59,000 years in the past, more than 80,000 years after the African Eve. The chronology may prove too young.
>
> Molecular biology provides a convincing general framework for the origins of *Homo sapiens* deep in Africa before 150,000 years ago.

from a common ancestral population that dates to somewhere around 170,000 years ago. As far as we can tell, no modern humans lived outside Africa until around 59,000 years ago.[7]

The African Eve is a fictional person, a product of molecular biology, which has used mitochondrial DNA to show that all of us, wherever we live, are ultimately of African descent. If such a person existed, she would have been dark haired and black skinned, a member of a small hunting band, and strong enough to tear apart human flesh with her hands and carry heavy loads. She was not the only woman on earth, nor was she the most attractive, nor even the one with the most children. But she was so fruitful that a certain set of genes passed from her into every living being on earth today.

Molecular biology has traced the ancestry of the Cro-Magnons deep into tropical Africa, into the territory of the hypothetical African Eve. The genetic framework is plausible, but what can archaeology tell us to support it?

WHEN BERNARD VANDERMEERSCH excavated the Qafzeh cave, in Israel, he originally estimated that the modern-looking human burials from there were about forty thousand years old. A year after the publication of the Cann paper, French and Israeli scientists dated burned stone flakes from the site using the thermoluminescence method and came out with dates of about ninety-two thousand years ago.[8] In one fell swoop, the chronology of *Homo sapiens* in the Near East moved back fifty thousand years. Recently, the Skhul burials were redated to virtually the same time period, much earlier than the original estimate of forty thousand-years ago.

With both mtDNA and the new Qafzeh dates on the table, a new generation of theories argued that *Homo sapiens* had originated in tropical Africa some 150,000 years ago and had spread into the Near East by about 100,000 years before the present.[9] Why would such population movements have occurred? At the time when the Qafzeh people lived, goes the argument, the Near Eastern coast was essentially part of northeast Africa, better watered than today and a place that was readily accessible from both Arabia and the Nile Valley. Without question, then, there were irregular human population movements across this vast, semiarid area about 100,000 years ago, perhaps through now dried-up and buried river valleys in the Sahara, which may have brought *Homo sapiens* groups into western Asia, either as occasional visitors or as permanent residents. There were never many migrants; they probably hunted and foraged in almost the same way as their Neanderthal contemporaries, and their technology would have been virtually identical to that of their neighbors. In appearance, they would have displayed a mingling of more archaic and modern features, with the reduced browridges and modern vocal tracts of the Skhul people. Effectively, they were moderns, but their intellectual abilities were basically those of earlier humans, even if their stone tool manufacturing was a little more efficient and flexible than that of their predecessors. The cognitive revolution that ultimately turned premoderns into moderns had not yet occurred. Interestingly, one modern human buried at es-Skhul lay with a red deer antler, while one at Kebara cave, also in the Mt. Carmel range and close to es-Skhul, had a boar jaw in his or her grave. There were perforated marine shells with the Kebara graves, while red ocher came from sources nineteen miles (thirty kilometers) away. It is as if the small modern populations had more extensive social networks than their predecessors.

What do we know about early *Homo sapiens* populations in Africa at the time of Qafzeh and before? Our information is sketchy at best. Judging from a handful of African fossils dating to between 300,000 and 50,000 years ago, the evolutionary process was a slow one that took as long as half a million years. During these millennia, Africans must have displayed a considerable range of variation, with, perhaps, the final "modernization" of the skull and the enlargement of the brain, with its fully modern intellectual capabilities, occurring relatively late in the process. A few discoveries provide clues. A tall, well-built male with a broad forehead and thinning browridges lived at Omo Kibish, in Ethiopia, about 195,000 years ago. In 1997, the Ethiopian paleanthropologist Yohannes Haile-Selassie unearthed three fully modern 160,000-year-old skulls, one of them from a child, at Herto, also in Ethiopia. Herto shows that by then the anatomical development of *Homo sapiens* had run its course, for the skulls are virtually identical to those of modern people.[10]

Omo Kibish and Herto are conclusive proof that modern-looking humans flourished in tropical Africa long before 100,000 years ago. Few in numbers, they lived in small, extremely isolated hunting bands. We know of this isolation from the National Geographic Society's Genographic Project, a major effort to track human migrations using DNA. As part of this ambitious research, a consortium of geneticists constructed a matrilineal family tree (passing through the female line) of 624 complete mtDNA genomes from living Africans.[11] They concentrated much of their effort on the Khoi and San peoples of southern Africa, because they are surviving representatives of ancient hunter-gatherer traditions—people with a slender, light build, quite different from the squat, cold-adapted build of European Neanderthals. Their paternal and maternal lineages are along the deepest branches known among modern humans. The consortium believes that a major split in the human mtDNA tree occurred between 140,000 and 210,000 years ago, perhaps caused by genetic drift resulting from the persistent isolation of small human populations at the time. At this point, small *Homo sapiens* populations in East Africa and southern Africa became isolated from one another for about 70,000 years or so, until around 70,000 years ago.

Africa is an enormous continent, where early-human populations were always thin on the ground, even during periods of ample rainfall.

Sparse numbers is one thing, extreme isolation quite another, so the cause must have been a climatic one. What sub-Saharan Africa was like during those critical fifty millennia remains a question mark, for until recently the only information on climate change over 150,000 years ago came from deep-sea cores, not from on land. Then as now, monsoon rains and El Niños played a major role in bringing ample rainfall or drought to the lands south of the Sahara where the first modern-human populations thrived. Or at least they appear to have thrived until a series of megadroughts descended on East Africa and southern Africa after 135,000 years ago.

We knew nothing of these droughts until a team of researchers laboriously transported a drill rig to the waters of Lakes Malawi and Tanganyika, in East Africa, two of the world's oldest and deepest lakes. Lake Malawi is an inland sea by any standards.[12] I remember being caught out in deep water when a sudden storm blew up out of a clear sky. The wind gusted to thirty knots or more. Within minutes, short, steep waves threatened to overwhelm our small, open launch. Fortunately, the skipper had been caught out before. He turned the bow into the wind and waves and kept the boat stationary with the engine. The violent motion made us all seasick. I wished I had never been born. As the sun set, the wind switched off as abruptly as it had blown up, and we made our way thankfully to shore.

Had you told me at the time that Lake Malawi virtually dried up 70,000 years ago, I would have laughed at you. But it did. One of the cores drilled out of the northern depths of the lake penetrated an area that had mainly been submerged for the past million years. The core documented great climatic variability after 135,000 years ago, marked by periods of extreme, prolonged drought, with shorter intervals of higher rainfall about every 11,000 years. There were periods of severe droughts between about 135,000 and 75,000 years ago, which reduced the lake's water level by at least 95 percent. The level of aridity was almost unimaginable: 75,000 years ago, 341-mile-(550-kilometer)-long Lake Malawi was a couple of insignificant pools no more than 6 miles (10 kilometers) across and 656 feet (200 meters) deep. Only after about 70,000 years ago did wetter, more stable conditions return, at which time lake levels rose dramatically. Mitochondrial lineages hint that modern-human populations in both East Africa and southern Africa now expanded their ranges as their isolation ended and their numbers increased. There were very few of them indeed. They were the lucky survivors of an epochal natural disaster on the other side of the world that almost sounded the death knell of *Homo sapiens*.

SOUTHEAST ASIA, ABOUT 73,500 years ago. The even tenor of life unfolds day after day in the dense forest. Daily routines change little with the passage of the seasons. Two or three men armed with bamboo spears stalk monkeys and small game among the trees and at the edge of clearings. Their hunting territory is small, defined as much by edible plant foods, collected for the most part by the women, as by animals, and by encampments little more than a cluster of grass or palm frond shelters, familiar stopping points in a humid, tropical world—until the ash comes.

The deluge comes without warning, gray, menacing, and enveloping. Choking ash and dust descend from the heavens, mantling forest and open country alike. The siege continues for at least six days and nights. Animals and humans cough and wheeze in the suffocating ash and heat. There is no escape. Everywhere, the landscape turns gray and lifeless. Animals and people die, powerless to evade the sudden attack. Only a few hunting bands survive, those who are lucky enough to be close to caves and natural overhangs, or who live in rare sheltered enclaves, where vagaries of topography and prevailing wind protect them. Even those who emerge unscathed from the onslaught suffer from hunger. Inevitably, many of them also perish in a now-devastated tropical world, from Southeast Asia and South Asia across a broad swath of the tropical world as far as Arabia . . .

Mount Toba, a large volcano on Sumatra and part of the notorious Pacific "Ring of Fire," had exploded in one of the greatest eruptions of all time.[13]

The blast was stupendous: 670 cubic miles (2,800 cubic kilometers) of the mountain vanished into space. The lava flows from the eruption covered over 7,700 square miles (20,000 square kilometers), reaching the southwestern and northern shores of northern Sumatra. Further hundreds of square miles of solid rock fractured into vast clouds of volcanic ash that rose over 20 miles (32.5 kilometers) into the atmosphere. Millions of tons of sulfur gas reached the stratosphere and lingered there for years. When the eruption subsided, only a huge crater remained, now the world's largest volcanic-crater lake—62 miles (100 kilometers) long, 18.6 miles (30 kilometers) wide, and up to 1,666 feet (505 meters) deep.

The Toba eruption was the greatest volcanic event of the past twenty-three million years. To put it in perspective, another Southeast Asian volcano, Mount Tambora, erupted in 1815. Tambora's ash clouds mantled the sun over much of the world and caused the famous "Year Without a Summer" in 1816.[14] Swiss farmers starved. The cold along the Swiss lakes forced the vacationing poets Lord Byron and Percy Bysshe Shelley and the latter's soon-to-be wife, Mary Godwin, to stay indoors. They passed the time by telling fanciful tales. Mary created Frankenstein. Tambora's temper tantrum was forty times *smaller* than Toba's. The notorious Krakatau explosion of 1883, heard hundreds of miles away, was even smaller—and that was a monstrous catastrophe.

How do we know of such a massive explosion when all that remains is a large volcanic lake? It has taken years of patient mapping and satellite observation to plot the extent of the eruption, and especially of the ash that fell on so much of Asia. Large-scale maps show that Toba's ash clouds spread to the north and west. Volcanic debris on land and on the seabed covered a huge area of the tropical world—the northeastern Arabian Sea, much of the Indian Ocean as far south as fourteen degrees south of the equator, northern India and Bangladesh up to twenty-five degrees north of the equator, most of island and mainland Southeast Asia, and the South China Sea. The thickness of the ash layer varied from place to place, sometimes a few inches, more often much deeper. Some ash deposits in Indian lakes are up to nine feet (three meters) thick. Huge areas of the tropical world, whether forest, savanna, or semiarid lands, lay inert under Toba's ash.

Toba was the ultimate in human disasters. Perhaps as few as ten thousand people, maybe fewer, survived the short- and long-term consequences of the cataclysm, most of them in cooler environments far from the blast. Humanity nearly became extinct.

The ash was just the first chapter of the disaster. The massive volcanic dust clouds did not remain in suspension that long. However, the sulfur gas from the eruption formed a long-lasting stratospheric haze that reflected sunlight and triggered severe, more prolonged global cooling. Almost immediately, the temperature of the South China Sea dropped by 1.8 degrees Fahrenheit (1 degrees Celsius). One Greenland ice core displays a massive sulfur peak lasting about six years at seventy-one thousand years ago. The core also documents a two-hundred-year period of unusually high amounts of windblown dust, probably because of decreased vegetational cover. The temperature over Greenland may have dropped as much as 11 degrees Fahrenheit (6 degrees Celsius). Some of the lowest temperatures of the late Ice Age coincided with the Toba eruption and lasted for as long as two thousand years.

The sulfur gas rose exceptionally high into the atmosphere, so the effects endured, to the point that anecdotal accounts of the effects of Tambora or Krakatau, which lasted a relatively short time, are irrelevant. For years after Toba, the sun shone faintly behind veils of dust, and temperatures fell. The effects were especially severe in the tropics. Tropical forests are extremely vulnerable to cooling, for seed banks are often limited and aboveground plant tissues die rapidly when exposed to even a few days of low temperatures. Temperate forests can also suffer badly from summer cold, especially deciduous trees, which lose their sprouting foliage almost immediately. The sudden cold may have also coincided with significant cooling of the southwest Pacific and weakening of the trade wind circulation. As a result, long-term El Niño–like conditions may have contributed to the megadrought conditions already prevailing over much of Africa.

The human cost was enormous. Thousands of people perished from starvation, including, almost certainly, most, if not all, of the *Homo sapiens* population of the Near East. Being of African origin and adapted to warmer temperatures, they were probably unable to cope with drastically lower temperatures. Geneticists theorize that a severe population bottleneck ensued among humans over a wide area about seventy thousand years ago.

Population bottlenecks involve substantial reductions in the size of a population that lead to the extinction of many genetic lineages within it. Such bottlenecks have affected many animal species. For example, the endangered Yangtze freshwater dolphin, whose numbers are now under a hundred, is currently in such a bottleneck, and may even be extinct, because of overfishing and pollution. Endangered species that don't vanish can expand once again later on, but with limited genetic diversity. The lack of genetic diversity among humans results from a series of bottlenecks over the past million years or more, the last of which was the one believed to have taken place about seventy thousand years ago. We have remarkably little genetic diversity compared with our closest living relatives, chimpanzees; indeed, there is more genetic diversity between individual members of chimpanzee troops in West Africa than among all living humans on earth.

The Mount Toba disaster decimated animal and plant communities over much of the tropics. Africa's human population had grown substantially through the last interglacial, which had brought warm temperatures and higher rainfall to the Near East, the climatic conditions that had brought a few *Homo sapiens* groups to western Asia, to Qafzeh and es-Skhul. South of the Sahara, the extreme aridity of the megadroughts had already decimated human groups. Mount Toba added exceptional cold and even drier conditions to the climatic equation. Africa's population tumbled again as arid conditions and the cold brought famine. Geneticists estimate that the African population declined to a total of between four thousand and ten thousand females of reproductive age. Add to this an average of two children, a male mate, and a small number of elders beyond reproductive age, and you have a population that would not fill the average Big Ten college football stadium.[15] This, by any standard, was a narrow bottleneck, especially when you reflect that there are about a billion reproducing females on earth today.

The survivors dwelled in refuge areas in East Africa and southern Africa, living in complete isolation from the wider world, and usually from neighboring groups. The effects of genetic drift may have kicked in, the accumulation of random changes in gene frequencies in isolated populations that may have compounded over time. The genetic bottleneck lasted for some twenty thousand years. These were some of the most critical millennia in human history, for it was during this period that *Homo sapiens* acquired their full cognitive powers. Once the bottleneck ended and rainfall improved, Africa's population grew rapidly. *Homo sapiens* once again moved out of the tropics, but this time with a difference. They possessed all the awesome mental abilities of modern humanity.

How DID WE acquire the cognitive skills we have today? This really is the question of questions, one that baffles researchers, for it involves a venture into uncertain territory. Language and thoughts, symbolic meanings and emotions, are transitory phenomena, which appear to archaeologists only indirectly, for we gaze through a clouded lens into the realm of the ancient intangible. This reality makes it extraordinarily difficult for us to pin down the stirrings of modern human cognition, except in the most general terms and with indirect clues. These signs, such as they are, come from Africa, at about the time of, and after, the Mount Toba cataclysm, and involve the exotica of stone tool technology.

Yes, stone tools again: by about three hundred thousand years ago, and perhaps even earlier, long before Neanderthals finally evolved in Europe, some African groups were making large, thick stone blades (a blade is a flake that is longer than its width), for the most part from local materials, which were conveniently at hand. They apparently made no effort to obtain fine-grained toolmaking stone from other locations. This technology remained in use for thousands of years. Then, about seventy thousand years ago, during the cold, dry interval coinciding with the Mount Toba disaster, some South African hunter-gatherers developed more refined stone artifacts. Instead of making large spear points, the stoneworkers struck off thin blades from cores of fine-grained rock that did not occur locally. Many of these blanks then received blunted backs before being mounted as barbs on spear shafts or on wooden handles for other purposes. These refinements were episodic; that is, they never became permanent, long-term parts of human tool kits, being replaced with older, larger flakelike artifacts during subsequent warmer millennia.

Whenever the new technologies came into use, the size of the tools they yielded was considerably smaller, as if there was a fundamental change in hunting methods. There was more to it than merely episodic changes in stone artifacts.[16] At the Blombos cave, near the southern tip of South Africa, the inhabitants of about seventy-three thousand years ago ate a broad array of foods, including not only game and plants but also fish and mollusks. They hunted with finely made, stone-tipped spears, but they also used delicate bone points, lethally sharp projectile heads, especially when smeared with vegetable poisons, which may have been used for the first time there. The nearby Sibudu cave received visitors about sixty-one thousand years ago, and they left behind a bone projectile point.[17] They also manufactured small tools with blunted backs, perhaps as barbs for use on hunting weapons. Many of them show distinctive fractures that are characteristic of breaks caused by a sharp impact. The wear patterns on the working edges are consistent with those on artifacts used as barbs or mounted heads.

The Sibudu hunters preyed not on larger antelope such as eland, but, for the most part, on the duiker, a small, shy antelope that rarely weighs more than eleven pounds (five kilograms) and thrives in forested environments. One can imagine a hunter stalking his solitary prey as it browses on the stunted, brown-yellow grass among a grove of evergreen trees. The duiker flaps its short tail as flies swarm around it. Intent, its pursuer ignores the buzzing insects, light spear at the ready, creeping ever closer from downwind, using the tree trunks as cover. The duiker feeds in the shade to escape the midday heat, looking up quickly at the slightest sound. Time after time, the hunter freezes absolutely still for minutes, until he can move almost imperceptibly closer. Finally, he's in range, virtually invisible behind a strategic tree. A quick cast hits the duiker in the flank. The razor-sharp stone barbs of his spear cause blood to spurt from the wound as the shaft breaks off, leaving the head behind. As the duiker rears up and tries to flee, his pursuer quickly fits another point to his spear, runs forward, and thrusts again at the staggering beast . . .

Could, then, the making of bone points and the use of smaller stone tools imply a major change in hunting methods? Heavy spears, whether stone tipped or not, are effective against larger animals and offer the only way of killing them without the help of vegetable poisons. In the arid, open environments of southern Africa seventy thousand years ago, lightweight weapons used against smaller animals like the duiker would have been far more effective, especially if smeared with vegetable poisons. Such spears would have also had a greater range, which would have helped in open country, where close-range stalking would have been difficult.

Smaller stone tools and bone-tipped weapons ebbed and flowed through the African world of sixty thousand to seventy thousand years ago, eventually vanishing, only to reappear thousands of years later. We should not be surprised by this, for the severe aridity and climatic changes of those millennia would have caused sparse and isolated hunting groups to adapt constantly to new conditions.

Another change occurred at about the same time. Stoneworkers became highly selective about toolmaking rocks.[18] Those who manufactured stone barbs and other smaller artifacts relied on fine-grained rocks such as obsidian (a volcanic glass), often obtained from distances of 25 miles (40 kilometers) away or more. At Prolonged Drift, in the Kenya Rift Valley, a temporary hunting camp dating to around seventy thousand years ago, the inhabitants obtained half of their obsidian from 30 miles (50 kilometers) away and 40 percent from 25 miles (40 kilometers) away, this despite there being sources within 19 miles (30.5 kilometers) or less of the site. At the time, the environment was cold and dry, which may have forced the people to range widely over large hunting territories, hunting opportunistically for food as well as toolmaking stone.

Much of Africa was now dry and cool, with shrunken forests and human populations isolated in sheltered enclaves. If the South African sites of the time are any guide, some groups lived by coastlines, where, in addition to game and plant foods, they relied on fish and mollusks, some of the earliest known fishing in the world. Wherever people lived, they must have had to cope with more open, drier environments, where game, plant foods, and water supplies were scattered and sometimes clustered over large areas. No longer could they rely on the ancient strategies of moving across small, familiar territories, their movements attuned to the seasons of plant foods and their prey. Now food supplies and water holes shifted and changed with the seasons and much more irregular rainfall.

Each hunting band, living as it was in isolation, had to depend on information acquired from over the horizon, from neighbors moving across nearby lands. Everyone was at risk of starvation, so much depended on sharing intelligence—about game, water, and ripening nuts. Sharing information may have stimulated the development of fluent speech, of sophisticated communication. We have, of course, no idea how intelligence gathering operated, but it may have been somewhat like the now much-elaborated but ancient *hxaro* system of partnerships with neighbors near and far used by living San of the Kalahari Desert, which is based on gift giving and reciprocal obligation.[19] This helps greatly in reducing risk in unpredictable, arid environments. I've watched San hunters sitting apparently idle in the shade talking for hours on end. I soon learned that their seeming leisure was in fact vital to survival—a time when people who foraged over a wide area shared information on food and water.

A number of archaeologists believe that the first such intelligence systems, based on kin ties and reciprocal obligation, appeared around seventy thousand years ago in response to prolonged drought and the ravages wrought by the Mount Toba eruption. They also wonder if the medium for gift exchange was fine-grained toolmaking stone, or perhaps the light weapons made with it. The problem with such gifts is that they would have gotten used up, been broken, or worn out. So the gift exchanges may have involved more durable items as well, such as ostrich eggshell beads, today still a universal part of *hxaro* in the Kalahari Desert. Such gifts could have been worn and displayed, as symbols of mutual dependence between gift partners. They would also have been visual symbols of social solidarity, of alliances and partnerships between people living near and far that would have been highly adaptive in high-risk environments like those brought by Toba. Over time such networks revolutionized human social behavior and communication. Above all, they would have involved the exchange of intelligence, perhaps planning and organizing a hunt or other activities. As part of this process, the exchange of ideas, there may have been the first steps toward developing specialized tool kits that were highly portable and easily replaced or repaired.

The harsh environmental conditions after the Toba disaster would have fostered strong pressures for cooperation over longer distances, even entire regions. Such cooperation would have come from expanding social networks far beyond the limited contacts between neighbors. Increased social activity may have triggered occasional meetings and ceremonies, during which kin and gift partners would exchange information and discuss details of hunts. Survival depended on cumulative experience, on information passed from one generation to the next through example, and now, perhaps, through articulate speech. Such experience would have surrounded not only human relationships but also the complex links between people and their prey and among the living, the dead, and those still to be born. There was a new realm of symbolic meanings, which thrived in a world of partnerships between humans and their surroundings, expressed both verbally and, for the first time, in artistic impression. The Blombos cave has yielded striated red ocher, as well as the earliest known ornaments: perforated *Nassarius* seashells. (Similar perforated shells have recently been found in a cave in Morocco.) The world's first art, if you wish, but these humble objects are the first signs that *Homo sapiens* was developing the cognitive abilities that made the Africans of about seventy thousand years ago like us.

These developments came at a time of stress and reduced population, when only a few thousand people lived south of the Sahara and when Africa was cut off from the outside world by intense drought. Then, after seventy thousand years ago, Africa's population began to grow once again as the cold receded. Small numbers of people, with all the cognitive abilities of modern humans, now moved out of Africa into Asia and lands beyond, perhaps between seventy thousand and fifty thousand years before the present.

How do we know about this later, and definitive, migration? The Y chromosome provides a male family tree that suggests that no modern humans lived outside Africa until about fifty-nine thousand years ago.

This may have been about the time when fully modern *Homo sapiens* finally spread out of Africa onto other continents, just at the time when human populations were swelling after the Mount Toba catastrophe.

RESEARCH IN MOLECULAR biology moves faster than in archaeology, where a single site can take years to excavate. For every month's digging, there's at least three months' laboratory work to follow. So far, the putative routes followed by *Homo sapiens* out of Africa remain unknown, the signature of human passage even under ideal circumstances being faint at best.[20] Only a few thousand people, moving in small bands, would have crossed into Asia, perhaps along the Nile Valley, down then-flowing Saharan watercourses, or across a Red Sea whose levels were much lower during the late Ice Age. Imperceptibly, in tiny numbers, the African ancestors of the Cro-Magnons expanded out of their tropical homeland at a time when human populations were on the rebound. These were people who were accustomed to maintaining connections with others over long distances, to whom ideas of reciprocity and exchanging information were routine. They would cooperate with others, had efficient weaponry, and could not only tackle animals large and small but also exploit all kinds of other foods, especially plants. Above all, they were planners and thinkers, people who valued their relationships with both the living and those who had gone before, and with the animals, plants, and landmarks that defined their world. They were at one with their surroundings, no longer another predator among predators, but true human beings—ourselves.

Small in number, widely dispersed over inhospitable terrain—these anatomically modern people from Africa and then the Near East were the primordial ancestors of the Cro-Magnons. Within a remarkably short time, some of their descendants moved out of the Near East into Eurasia and Europe—to a completely different world.

CHAPTER 6

Great Mobility

THE NILE VALLEY, SPRING, fifty thousand years ago. The blue waters of the river tumble between dark boulders, shining black in the strengthening morning sun. Here, the fast-flowing Nile weaves through a narrow valley, the desert pressing close on either bank. Only a narrow strip of green separates water from windswept sand, just enough for a small band to camp among rocks that shelter them from the north wind. The rapids debouch into a deep pool where the translucent water lingers placidly on its way downstream. Two men stand on boulders, spears in hand. They peer into the calm, shallower water near the bank. They know that catfish linger close to the surface here, where they feed in the mud and sand. One of the fishermen raises his bone-tipped spear, aiming at a shadow on the muddy bottom. The dark shape moves sluggishly. He casts rapidly, with unerring aim. The water boils and swirls as the catfish tries to escape. A deft flick with the spear, and the fish is ashore, flopping helplessly in the hot sun. Seconds later, the fisherman kills his prey with a quick blow from a wooden club, then moves the carcass into the shade. The pursuit resumes without delay. By the time the sun is overhead, the two men have caught enough for the evening meal.

Meanwhile, the women pluck bunches of nut grass among the reeds along the river, keeping a watchful eye out for crocodiles. They jerk mature tubers from the soil, then carry them back to camp, where they pound and grind the fibrous roots on flat-topped stones before leaching them in water to remove bitter toxins. When the men return with their catch, everyone guts the fish and cuts the flesh into narrow strips. Some they will cook over the fire. The rest they hang up on wooden racks to dry in the sun.

How do we know that hunting bands fished along the Nile fifty thousand years ago? Unfortunately, we don't, but the fishing scene in the preceding paragraphs most likely reflects what actually happened there as small numbers of modern humans spread out of Africa. The Nile Valley is one of several logical routes to the north and east. So are long dried-up watercourses in the depths of the Sahara, if the desert was slightly wetter at the time. Another theory brings hunting bands across the Red Sea at a time of much lower sea levels into Arabia and then northward. Whichever route you favor, your guess is as good as mine. (Information on climate change during the late Ice Age appears in figure 6.1.)

Looking for signs of the irregular population movements that brought *Homo sapiens* out of Africa about fifty thousand years ago is like looking for the proverbial needle in a haystack—or even worse, because the needle is almost invisible. These were people on the move, bands adapted to dry, open terrain, where hunting territories often covered hundreds of square miles. Rare were locations where a group could stay in one place for any length of time. Nearly everywhere, campsites were fleeting at best, occupied, perhaps, for a few days or weeks. When the inhabitants moved on, they left behind some seed husks and other plant remains, a few broken animal bones, and perhaps a collapsed grass shelter or a scatter of stone flakes and tools. Within a few months, anything organic vanished, leaving a scant few grinders and stone fragments for the inquisitive archaeologist. After the passing of fifty millennia, the signature they left on the ground is virtually invisible. I've spent days looking for such ephemeral sites in central Africa—day after day of tracking tiny bunches of humanly struck stone flakes, often revealed in the entrances of fresh rabbit holes, or on recently burned ground. This is archaeology at its hottest and dullest, even when the ground is paved with stone tools, as it is, for example, along long stretches of the Middle Zambezi Valley. Only rarely do you find a finished artifact or a hearth. Most of the time, it's just waste flakes and cores, the debris of a transitory passing. If my experience is any guide, studying the details of stone artifacts made in open camps in the earliest moderns' homeland and along the Mediterranean shore is a near-fruitless

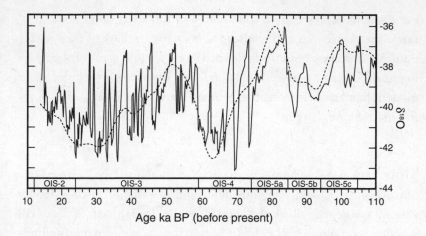

LATE ICE AGE CLIMATE CHANGE

Years ago	Event	Conditions
12,000	Onset of Holocene: modern times	Warming
13,000	Younger Dryas cold event	Near-glacial
15,000	Oscillating climate.	Irregular warming
25,000	Last Glacial Maximum	Very cold
30,000	Gradual cooling with some warm intervals.	Neanderthals extinct
45,000	Irregular climatic shifts.	Some intensely cold moments.
60,000	Brief warm interval	Modern humans arrive
74,000	First intense cold period.	Neanderthals
115,000	Beginning of slow climatic deterioration	
Before 115,000	Last interglacial—warm/temperate	

Figure 6.1 *Climate change during the late Ice Age. (top) A diagram of the zigzag climatic shifts from 110,000 to 10,000 years ago, as reconstructed from a variety of sources, especially Greenland ice cores. The relevant Oxygen Isotope Stages (OIS) are shown. (bottom) A summary of major events.*

exercise, unless preservation conditions are truly exceptional—and that's rarely the case in semiarid landscapes. It's naive to think of their makers dutifully scattering beautifully finished stone tools as signs of their passing from Africa into western Asia and then Europe. All we know is that modern humans were living over a broad area of the Near East by forty-five thousand years ago.

MOST LATE ICE Age human settlement in the Near East centered on a narrow strip of oak, terebinth, and pine woodland on the Mediterranean coast and along the flanks of the Jordan Rift Valley. This wooded terrain provided rich nut harvests, a variety of game, and relatively reliable water supplies. A small corridor of landscape confined by deserts, the Mediterranean, and mountain ranges, the varied terrain boasted numerous caves and rock shelters, which both Neanderthals and modern humans visited over long periods of time. The human population of this highly strategic region can never have been large; one estimate is about sixty-four hundred people—and that may be too high.[1] How many of them were moderns as opposed to Neanderthals is a matter of debate, as are the relationships between them.

There were, of course, many similarities between the two groups in terms of prey they hunted and so on, which were reflections of living in very similar environments. But after about forty-five thousand years ago, the moderns settled without hesitation in arider habitats of low productivity where Neanderthals rarely ventured. They also hunted a much broader range of animals, among them rabbits, rodents, birds, and tortoises, which gave them a more stable subsistence base. Their weaponry was lighter, and they used small stone points on their spears and, occasionally, bone spearheads (figure 6.2). These were highly adaptable hunter-gatherers, as at home in deserts as they were in better-watered landscapes. It is as if there was a significant change in human behavior, a break from the ways of earlier times.

How and why did such a break take place? What would cause one population to replace another? The obvious place to start is with climate change, which has become the new player on the early-modern

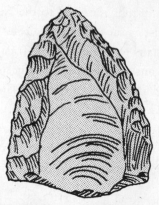

Figure 6.2 *Emiran point, a light-weight stone spearhead used widely over the Near East, c. forty-five thousand years ago.*

historical stage in recent years. Speleothems (stalagmites and stalactites) from two caves in Israel provide direct evidence of rainfall fluctuations in the core area of human settlement in the Near East. A major drop in both rainfall and temperatures occurred between seventy-five thousand and seventy thousand years ago, the time of the first cold maximum of the last glaciation, and a second one took place between about forty-seven thousand and forty-two thousand years before the present. During these same periods, pollen diagrams tell us, tree cover thinned dramatically and steppe-desert vegetation spread. Many smaller warm-loving animals vanished, to be replaced by colder-loving forms.[2]

At the time of the cold snap around seventy thousand years ago, small numbers of modern humans from the earlier African migration still dwelled in the Near East. They vanish abruptly in the prehistorical record, at the moment when Mount Toba blew into space and a "volcanic winter" descended on the world. The Near East soon became colder and drier. The Sahara and Arabian deserts expanded considerably, isolating the tiny numbers of moderns in a small enclave with sparse permanent water supplies and small patches of woodland. Between seventy-five thousand and seventy thousand years ago, the first *Homo sapiens* groups to move out of Africa apparently dwindled into extinction.

Warmer, wetter conditions returned to the Near East after seventy thousand years ago, just as they did to tropical Africa. You might have expected modern humans to move out of Africa once again, but they did not, probably because of the catastrophic population bottleneck that had reduced their numbers to a few thousand people. There were no pressures to expand northward when Africa was virtually deserted. Instead, Neanderthal bands reoccupied wide areas of the region. Whether they moved from elsewhere into a now-uninhabited landscape or had remained there all the time and simply increased in numbers is still an open question.

Some twenty thousand years later, a five-thousand-year cooling spell caused the Neanderthals to contract into much smaller territories near the Mediterranean coast. They now spent longer periods of time at settlements along the coastal lowlands. Hearths became larger; people consumed less-favored large animals, ate more plant foods, and used more specialized tools. Everyone seems to have worked harder just to survive, but there are no signs that they expanded to North Africa or the Nile Valley. For all these adjustments, the rhythm of their ancient lifeway remained fundamentally unchanged.

We don't know how many Neanderthals were living in the Near East when the moderns arrived, nor whether the two populations came into even sporadic contact. Food supplies were finite, water supplies often widely distributed, the carrying capacity of the landscape insufficient for more than very sparse human populations, making competition for food and water a virtual certainty. Did such competition mean that the newcomers pushed the primordial inhabitants aside into marginal territories where they eventually became extinct? Or had the Neanderthals in the Near East died out during the very dry millennia before *Homo sapiens* arrived? We don't know. After forty-five thousand years ago, no more Neanderthal fossils appear in the region's archaeological sites. They appear to have become extinct, perhaps by no coincidence, just as small numbers of modern Africans recolonized the drought-plagued lands.

Speleothems and other climatic proxies paint a picture of an arid landscape, with many fewer trees and more steppe-desert plants than tropical Africa. This does not appear to have deterred the newcomers, who came from a drought-plagued continent, from environments where both food supplies and water were scattered over large hunting territories. We know from burials that the new arrivals were tall and thin, with long legs, a biological adaptation to tropical conditions, which helped them spread widely over what was now a semiarid-to-arid steppe-desert world. Here they could thrive in environments that severely challenged even the most adept Neanderthals, who, as always, moved with the animals upon which they preyed, just as they had done for thousands of years.

Cooperation, planning, and intelligence: these were the trumpets that knocked down the proverbial walls of Jericho for *Homo sapiens* in the Near East. The newcomers were consummate networkers in ways that were alien to earlier humans. This gave them an edge that enabled them to exploit more diverse, much drier terrain. They were used to arid landscapes, to moving rapidly over large hunting territories, where intelligence gathering and contacts with others would have been of considerable importance. The moderns brought simple, proven technologies with them that were, at first, quite similar to those of the Neanderthals. There was, however, one significant difference: the newcomers possessed an ability to innovate and to develop new stone-flaking technologies and tools for their changing, often highly specific needs. Among them increasingly emphasized were thin, parallel-sided, and triangular stone blades, struck from carefully prepared conical and pyramid-shaped cores.

Tool kits were becoming lighter everywhere, even in the hands of Neanderthal stoneworkers, but the new arrivals refined blade making even further. As we saw in chapter 4, blade technology first came into use in tropical Africa as part of a trend toward lighter hunting weaponry that began sporadically about seventy thousand years ago as climatic conditions became wetter after the great megadroughts. Light weaponry armed with a lethal stone point was highly advantageous, especially if, as seems possible (but remains unproven), the hunter propelled his spear with a simple throwing stick. Spear throwers, described more fully in chapter 8, have the effect of extending the arm and increasing both the range and the velocity of a hunting weapon, an important consideration when stalking prey in open terrain with little cover and when pursuing small, nimble animals. Lighter projectiles thrown with telling accuracy at higher speeds would have been effective at much longer ranges, perhaps over two hundred feet (sixty-one meters) in skilled hands (see figure 8.2). Such weaponry would have provided a decisive edge in the chase, especially when combined with detachable spear points that could easily be replaced in the field. Jill Rhodes and Steven Churchill compared the anatomy of modern throwing athletes with those of a small sample of Neanderthals and Cro-Magnons. They found asymmetrical parallels between the upper-arm bones of modern athletes and late Ice Age moderns consistent with regular spear throwing. In contrast, the Neanderthals do not appear to have thrown spears regularly, using them more for thrusting.[3]

The new arrivals also enjoyed a broader diet. Wild plants of all kinds had been part of human subsistence in Africa for hundreds of thousands of years. Without question, such foods were still a significant part of the newcomers' diet, especially in areas of slightly higher rainfall, where fall nut harvests would have provided easily storable provisions . . .

Two groups of women meet by chance among the nut trees, although they could have got together by design, for they all know one another from previous encounters. After they exchange greetings, they sit in the shade during the heat of the day to eat some nuts and drink water from skin containers slung over their backs. It has been an unusually cold and dry year, so the nut harvest is late. Everyone has had to travel long distances for food. Even well-established water holes are running dry. The women laugh and gossip, but then the conversation turns serious, revolving, as it does, around water supplies. One of the women draws lines in the soil with a twig and uses tiny pebbles as rocks and nut groves to show where she found water a few days before. The others nod but report that when they were there yesterday, the spring was dry. The women decide that the best place to go lies over a nearby ridge, where they may be able to find some by digging into a dry riverbed that carries water for much of the year.

As the temperature cools, the foragers heft nut-laden skins onto their backs. They will meet again, as they have many times before. Back in one camp, the men return with a gazelle. They, too, have talked to their neighbors and have learned of abundant water perhaps a two days' walk away. No one keeps information to him- or herself, for they know that lives depend on shared intelligence in a dry landscape where conditions are always harsh . . .

Speed, mobility, constant innovation, planning, and ingenuity—these were the qualities that sustained *Homo sapiens* during the great diaspora that saw modern humans move out of Africa into the Near East. Tiny numbers of people moved over long distances with astounding speed, a reflection of both widespread aridity and a life that depended on constant mobility. We now find telling signs of social distinction, of individual and group identity, in the newcomers' settlements. Again, we enter the realm of the intangible, of long-vanished symbols like body paint, distinctive hairstyles, and headdresses and colored feathers that would establish one's kin and band affiliation at a glance. As always, all we have are durable objects: pendants and beads appear, including perforated marine shells and teeth, as well as ostrich eggshell beads, some traded over distances of 50 miles (80.5 kilometers) or more.

Small in number, widely dispersed over inhospitable terrain—these were the primordial ancestors of the Cro-Magnons. Within a few millennia, some of their descendants moved out of the Near East into Eurasia and Europe—to a completely different world.

THEY ARRIVED IN the north in small numbers, group by group, accustomed to open terrain as well as woodland, to environments where food resources clustered in patches, especially in sheltered locales, near permanent water, and along game migration routes.[4] For thousands of years, they had foraged in landscapes where temperatures were often hot during the day and intensely cold at night. Now they moved, apparently effortlessly, into very different terrain, where subzero temperatures were commonplace every winter, the bestiary was generally larger and more diverse, and predators were especially common.

As was always the case, this was not a deliberate journey, undertaken as a conscious decision to occupy new terrain over the horizon, but a natural continuation of population movements that had begun in tropical

Africa thousands of years earlier. Life was always about movement, about finding water and food. Hunting territories expanded and contracted as game migrations faltered or water supplies became more plentiful a few miles away. Accidents, marriages, and quarrels caused bands to merge and split. It's no mystery why *Homo sapiens* covered enormous distances between forty thousand and fifty thousand years ago (figure 6.3). Most landscapes could only support a few people per square mile. Above all, the natural dynamics and flexibility of hunting and gathering in generally dry environments saw people covering long distances in remarkably short periods of time. They entered a seemingly empty world, where Neanderthal populations were rarely encountered. Many of the newcomers may never have seen their elusive, quiet neighbors.

Within ten thousand years or so of leaving Africa, small numbers of moderns had settled throughout much of southern Eurasia and Europe.

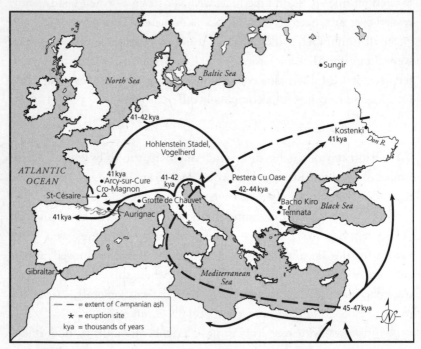

Figure 6.3 *The modern human settlement of Europe, showing the approximate extent of ash from the Campanian eruption.*

How and when they did so are two hotly debated questions. We are still looking for a needle in a haystack, but here the hay has been more thoroughly sifted than that in northeast Africa. The pickings are slim, as you might expect with only tiny numbers of people on the ground, compounded by difficult preservation conditions. All we have to work with are thin occupation levels in river valley soils in eastern Europe and in caves and rock shelters in the Danube Corridor and in the west. Fortunately, we have one defining moment: another massive volcanic eruption that scattered ash over much of eastern Europe thirty-nine thousand years ago.

The eruption of Mount Vesuvius in August A.D. 79 decimated the Roman towns of Herculaneum and Pompeii, on the Bay of Naples, and killed thousands of people. Vesuvius was nothing compared with the Campanian explosion, which devastated the same area thirty-nine thousand years (from present day) but had nothing to do with the former.[5] The violent upheaval collapsed a huge caldera spanning some 89 square miles (230 square kilometers) in an area that now lies under Naples and the northwestern portion of the Bay of Naples. Lava flows from the eruption extended over at least 50 miles (80 kilometers). A huge ash plume rose as high as 27 miles (44 kilometers) into the atmosphere and descended over a large area of the eastern Mediterranean and more than 1,550 miles (2,500 kilometers) northeast across the Balkans and into eastern Europe. Like the Toba eruption, the Campanian explosion must have caused widespread disruption to hunting territories and food supplies. Fortunately, people were sparse, which meant that a significant number survived during the ensuing cold event by moving into areas where game and edible plants could be found. The distinctive volcanic ash layer from the eruption over an enormous area forms an invaluable marker of known age in archaeological sites of the time. A classic volcanic winter followed the eruption.

The Campanian eruption was but a geological dwarf compared with the Toba cataclysm of 73,500 years ago, but its deposits are a godsend to the archaeologist chasing the Cro-Magnons, who colonized the north immediately before the cataclysm. Fortunately for science, their tools and food remains occasionally appear above and below the tephra (ash) layers from the explosion, which are like the filling of a chronological layer cake, even as far away as a place named Kostenki, by the Don river valley in the Russian Federation.

Tiny numbers of people settled on these eastern plains. They have left but the lightest imprints in the soil, even at long-occupied locations like Kostenki, a place famous for its mammoth bones and archaeological sites. (The Russian word *kost'* means "bone.") Most of the numerous Kostenki settlements, and those at nearby Borshchevo, lie at the mouths of large ravines cut into the west bank of the Don River, where springs come to the surface and game must have abounded. The first settlers occupied camps on the second terrace above the river, which lie below and above the weathered ash layers from the Campanian eruption.

Generations of archaeologists have worked at Kostenki, but none with the thoroughness of an international team of scientists who worked there in 2001–04. They used a battery of paleomagnetic, radiocarbon, and luminescence tests, as well as pollen and soil samples, to date these critically important layers.[6] It was cold when a small group of modern humans (identified at present from a single, highly diagnostic tooth, which may allow us to extrapolate beyond a single individual) settled briefly at Kostenki 14 with bone-tipped spears and other light weaponry between about forty-five thousand and forty thousand years ago. Temperatures were much warmer somewhat later when visitors came to the nearby Kostenki 17 site. Another single tooth shows that they were moderns who used both blade and bone tools, as well as perforated shell ornaments. There are other artifact scatters, too, which contain both blades and older artifact forms such as scrapers and triangular points.

Kostenki provides one archaeological footprint of the first settlers, perhaps as early as forty-five thousand years ago, but the tool kit from levels prior to the Campanian tephra is unique to the Don Valley, with no ties to contemporary artifact forms at sites in the Near East or further west. This should not surprise us, for the newcomers were highly mobile, few on the ground, and accustomed to adapting to local conditions with new hunting weapons and tools.

There are other traces of the newcomers farther west and along the Mediterranean, again lying below the Campanian tephra. The Temnata cave, in Bulgaria, contains blade tools and scrapers, as well as what are thought to be projectile heads similar to triangular points from the Near East. The occupation underlies the tephra and thus is more than forty thousand years old. Artifacts at the Bacho Kiro cave, 87 miles (140 kilometers) to the southwest, overlie earlier Neanderthal occupations and date to about forty-three thousand years ago.

We now know that these newcomers were anatomically modern people. The recently discovered Peştera cu Oase, "cave with bones," in southwestern Romania, lies in the Carpathian Mountains, a landscape of rugged peaks and deep valleys.[7] An underground stream created a network of galleries that were later blocked, so much so that the discoverers of the site accessed the corridors under water using scuba gear. The ancient entrances were eventually located with an electromagnetic transmitter, but the researchers did not open them. They carried out all the bones and other finds from excavations deep in the galleries, a remarkable achievement.

Peştera cu Oase was a place where cave bears hibernated. The galleries contained at least eight of their nests and the bones of over one hundred beasts. They hibernated there regularly until about 46,500 years ago. Later, wolves used the caves as a denning area. No humans ever dwelled at Peştera, but the jaw of a young adult and the skull of a fifteen-year-old came from the cave, dating to about 42,500 to 40,000 years ago, probably deposited there by sudden floods, which periodically ravaged the galleries. Both bone findings are from modern humans, but they display some archaic features, including very large third molars. These are the earliest known modern humans in Europe, with close resemblances to people from the Near East. They were, as biological anthropologist Erik Trinkaus puts it, "modern without being particularly modern."[8] (As these words were being written, the BBC was poised to air a TV program featuring a facial reconstruction of the Peştera adult.)

The pre-eruption footprints of the moderns extend into Greece and Italy, dating to around 42,500 years before the present. Traces of early occupation, often ill-defined mélanges of earlier technologies and more modern artifacts, appear across the southern half of Europe at about this time, as far west as northern Catalonia and along the Pyrenees to the Atlantic coast of northern Spain. Fortunately, the people who dispersed along Mediterranean shores made readily identified bladelets on specially prepared cores that produced larger blanks. Many sites in this time frame contain distinctive points, probably segments of multicomponent hunting weapons, the razorlike flints attached to wooden shafts with resin (for spearhead mounts, see figure 9.3). They appear in sufficient numbers to suggest some degree of cultural continuity over a huge area of southern Europe before the great eruption of thirty-nine thousand years ago.[9]

The great eruption may have been a deciding event in history at a time when the *Homo sapiens* population of Europe numbered no more than a few thousand people in isolated small bands, up from what may have been only a few hundred around forty-five thousand years ago. The Greenland ice record shows that they entered Europe during a warm period between about forty-four thousand and forty thousand years ago. We have no idea how the process unfolded, but I have a suspicion that it may have been somewhat like that of small numbers of much later Cro-Magnons, who expanded northward onto the northern European tundra after seventeen thousand years ago, as described in chapter 12. Perhaps the pioneers explored new terrain during the warmer months, following herds of horses, reindeer, and other prey. These were millennia of exploration and familiarization, of gaining intelligence about new landscapes, new sources of toolmaking stone, and other resources. Then the eruption and a cold event changed the dynamics of human life. Food shortages and the cold may have made contacts with others of prime importance, so much so that Cro-Magnon populations tended to concentrate in more-circumscribed areas. Isolation broke down; intelligence and contacts between different groups intensified; technological innovations flowered. In time, too, the innovation extended to social and ritual life: to art, music, and intricate beliefs that defined a harsh and ever-changing world. We can imagine the cautious first encounters between bands that rarely saw anyone else during their lifetimes: perhaps a chance sighting of people on a distant ridge, perhaps some hunters hearing of others by meeting survivors of a hunting accident. The exchanging of intelligence, a shared story, perhaps a joke, or a young man attracted to a girl—the sources of contact, then more-prolonged interaction, were open-ended and almost inevitable in a world where fluent speech was a catalyst, survival depended on cooperation, and symbolic meanings defined the landscape.

By thirty-nine thousand years ago, modern humans had lived in most of southern Europe for at least two thousand to three thousand years. Soon after the eruption, the same general tool kits, body ornaments, and social institutions flourished from the Don Valley to the Atlantic, a cultural tradition known to archaeologists as the Aurignacian (for artifacts, see figure 6.4). At Kostenki, Aurignacian tools appear above the Campanian tephra. By this time, between thirty-nine thousand and thirty thousand years ago, the makers of these Aurignacian tools killed large numbers of horses, which they probably drove in groups from the main Don Valley into smaller side ravines. Such culs-de-sac were natural traps for their prey, which could then be surrounded and killed. Freshly fractured horse bones abound in the kill sites, including largely intact back and feet bones, and these as well as cut and smashed bones testify to large-scale hunts conducted by people living in camps nearby. The hunting methods closely resemble those used thousands of years later in the same locations and at a Cro-Magnon kill site at Solutré, in France, described in chapter 10.[10]

WHY THE SEEMINGLY exotic name Aurignacian? It all began in France in 1858, in a small town named Aurignac in the foothills of the Pyrenees. A road worker found some human bones in a rabbit hole, dug them out, and exposed a cave with seventeen human skeletons. "Foul play," cried the local magistrate. He ordered the bones interred in the local cemetery. Two years later, Édouard Lartet of Les Eyzies fame excavated in the floor of the cave. He recovered flint and antler tools and the bones of extinct animals, which he proclaimed to be of "the remotest antiquity." Similar artifacts soon came to light during Lartet's excavations in Les Eyzies rock shelters. Fifty-two years after the Aurignac discoveries, archaeologist Henri Breuil defined an early Cro-Magnon culture and named it the Aurignacian in honor of the original discovery.[11]

Aurignacian, Gravettian, Solutrean, Magdalenian . . .

Archaeologists are great classifiers and subdividers, so much so that any story about the Cro-Magnons delves, inevitably and at least superficially, into arcane cultural names. It's no exaggeration to say that the history of the late Ice Age is wrought in stone and bone. Alas, very little else survives. Many of my undergraduate days were spent contemplating row after row of Stone Age artifacts, especially from France. Since that is where much Cro-Magnon research began, many distinctive artifacts and the cultures that begat them bear French names. Long-established archaeological practice names both after the site or location where they were first discovered—hence Magdalenian after the Magdalenian rock shelter at La Madeleine, near Les Eyzies.

For the purposes of this book, I use only six widely accepted cultural labels: Mousterian, Châtelperronian, Aurignacian, Gravettian, Solutrean, and Magdalenian. A brief visit with these terms at this stage in our story will allow me to use them without further definition in later chapters.

Archaeologists' names for different Cro-Magnon cultural traditions have deep roots in archaeological history, back to the days of Édouard Lartet and Henry Christy, in the 1860s. They excavated Le Moustier (Mousterian) and La Madeleine (Magdalenian), and Lartet dug the Aurignac cave (Aurignacian), just described. The later nineteenth century was a time of ardent evolutionary doctrines based on the notion that industrial civilization was the pinnacle of achievement to which all human societies aspired. Hardly surprising, the French prehistorian Gabriel de Mortillet wrote in 1867 of the "inevitable progress of Man," documented, so he claimed, in the orderly rows of stone tools from thick occupation levels in rock shelters at Les Eyzies. Mortillet and those who followed him developed a ladderlike succession of human cultures, starting with the Mousterian and ending with the Magdalenian. By the end of the century, the ladder had developed into a more complex evolutionary scheme in which the Aurignacian was followed by the short-lived Solutrean (after the horse kill site at Solutré) and then the Magdalenian.

Like all such cultural schemes, the ladder changed and was elaborated upon, especially in the hands of one of the immortals of Stone Age archaeology, the Abbé Henri Breuil, who achieved near-mythic status for his research on Cro-Magnon cave paintings in later life. A gifted artist, he copied Altamira and other sites long before color photography. His interpretations still influence our perceptions of the art, but his less well-known research on artifacts has cast just as long a shadow. In 1912, he published a conference paper on Cro-Magnon cultures in which he used distinctive artifact forms, "type fossils," as they are still called, to identify not only the major cultures but also subdivisions within them. Breuil worked almost entirely with western European sites, relying on his encyclopedic knowledge of sites and their occupation layers. For instance, two distinctive European type fossils, the split-based bone point and the steep-ended "carinated" scraper, marked the Aurignacian (see figure 6.4). He used occupation layers at Laugerie Haute and La Madeleine rock shelters, on the Vézère River, to subdivide the Magdalenian, the last and most elaborate of Cro-Magnon cultures, into no less than six subdivisions, the last two marked by single-barbed and double-barbed harpoons, respectively (figure 11.2).

Breuil's paper became archaeological writ for generations. His scheme survives today, albeit in much modified form. It is only recently, for example, that his six stages of the Magdalenian have come under fire. Others followed Breuil, among them Les Eyzies archaeologist Denis Peyrony, who argued that there were two parallel cultural traditions, the Aurignacian and the Perigordian, which later came together to form the Solutrean and Magdalenian.

Today's artifact classifications depend heavily on statistical methods, which were pioneered during the 1960s and 1970s for the

Neanderthals and Cro-Magnons by the French scholars François Bordes and Denise de Sonneville-Bordes. They developed minute classifications of stone artifacts using cumulative graphs of artifact percentages and other devices. Current methods are even more arcane but are yielding priceless information on the minutiae of Cro-Magnon technology. Nevertheless, the major cultural labels have persisted, except for the Perigordian, which has vanished.

The experts now commonly use the following as a very general framework:

Mousterian   A widely distributed technology of diverse human societies associated in general terms with the Neanderthals, circa one hundred thousand to thirty thousand years ago (see figure 4.1).

Châtelperronian   An enigmatic cultural entity in the West (see below), circa forty thousand years ago (see figure 6.5).

Aurignacian   Widely distributed over Europe and into the Near East, circa thirty-nine thousand to twenty-nine thousand years ago (see figure 6.4).

Gravettian   Widely distributed over Europe, with local variations, circa twenty-nine thousand to eighteen thousand years ago (see figure 9.3).

Solutrean   Northern Spain and southwest France, circa eighteen thousand to seventeen thousand years ago (see figure 11.2).

Magdalenian   Northern Spain to central Europe, circa seventeen thousand to eleven thousand years ago (see figures 11.3 and 11.5).

These are the general labels that I use in these pages. Each "culture" is marked by distinctive antler, bone, and stone artifacts, some of which are illustrated throughout the book. The deeper nuances of these Cro-Magnon cultures in artifact terms are really of more interest to specialist researchers than to general readers. When I refer to Aurignacian and Solutrean, I'm referring to different cultural phenomena marked by distinctive tool forms often found over wide areas, which changed over many millennia. However, a fundamental question remains: what do these archaeological terms mean in terms of what actually happened during the late Ice Age? Was there, in fact, a human culture that coincides with the archaeologist's Magdalenian, or did Cro-Magnon society simply evolve as dozens of tribal groups in ever more complex ways in a seamless continuum from forty thousand years ago or so until the end of the Ice Age and beyond? We will probably never know. The current archaeological labels, which are like a common scholarly language, acknowledge that there were significant and enduring cultural differences between widely scattered Cro-Magnon groups through time, even if we still do not fully understand what each term means in human terms. At present, they represent the best pathway for the general reader through a maze of contradictory artifact classifications.

Theirs was a simple but highly effective technology. Aurignacian stoneworkers crafted scrapers and burins, as well as unique *pointes d'Aurignac*, antler or sometimes bone projectile heads with a split at the base for mounting them on a spear shaft.[12] The tool inventory varies from site to site, but the basics remain the same: thick scraping tools made by flaking fine bladelets off the working edge, and often-quite-large blades with edge trimming or what appears to be deliberate notching, often called spokeshaves on the assumption that they were used for smoothing bone or wood (figure 6.4).

Unlike many such groupings, the Aurignacian is no local culture confined solely to southwest France. People fabricating similar artifacts, including split-based points, settled over an enormous area of Europe, from Kostenki in the east to the Atlantic in the west. Before World War II,

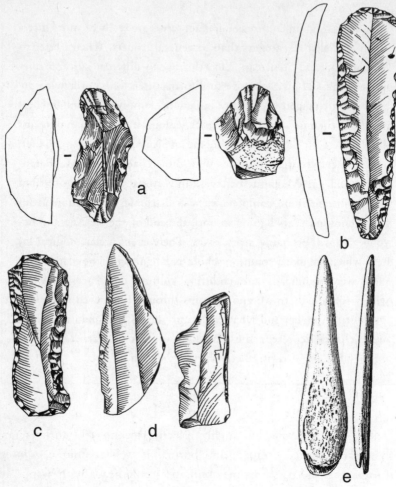

Figure 6.4 *Aurignacian tools. Aurignacian technology, based on blades, created large numbers of tools used as knives, woodworking implements, and chisels, as well as spear points. (a) "Carinated," steep-ended scrapers with steep working edges. (b) End scraper with working edge at its extremity. (c) Flake with sharpened edge, probably used for working skin and wood. (d) Burins with chisel edges. The small arrows locate the chisel edges. (e) Split base bone point, a so-called* pointe d'Aurignac. *After Jean-Luc Piel-Desruisseaux, 2007. Outils Préhistoriques (Paris: Dunod). Reproduced with permission.*

Dorothy Garrod identified Aurignacian occupation levels at Mugharet el-Wad, on Mount Carmel in the Near East. She boldly argued that the Aurignacians were the first modern humans to migrate into Europe, where they displaced the Neanderthals.[13] Even in her day, when much less was known about the late Ice Age, the wide distribution of Aurignacian sites through Europe was in noticeable contrast to the distribution of the sites of later Cro-Magnon societies. A remarkable continuity in Aurignacian technological practices extended over a distance of more than twenty-five hundred miles (around four thousand kilometers), from the Near East to northwestern Spain and as far north as England. No such technological, and presumably cultural, uniformity ever occurred again during the late Ice Age.

Was Garrod correct? Were the Aurignacians indeed the first settlers, or did earlier, still-shadowy groups arrive before them? We are lucky to have the bracketing tephra of the Campanian eruption, which dates scattered earlier modern-human settlements to before thirty-nine thousand years ago, whereas all known Aurignacian sites date to the time of the eruption or after it, the Aurignacian as a whole spanning about ten thousand years, until about twenty-nine thousand years ago.[14]

A concerted international effort surrounds the chronology of the Aurignacian and the so-called transitional industries that date to the millennia before the Campanian eruption. Some of these display mixtures of earlier Neanderthal technologies and the blade-making skills of the moderns. Do these mixtures imply that the new technology developed out of earlier traditions instead of being introduced from outside? There are many similarities between the old technology and the new, both of which are basically very simple. By the time moderns arrived, many Neanderthal groups were using more-diverse tool kits. Perhaps this was in response to changing climate, which may also account for the modernlike nature of some of their artifacts. All we know is that modern humans had arrived in Europe before the Campanian eruption and that after it Aurignacian culture flourished over an enormous area. Significantly, perhaps, the movement northward came during a period of warmer climate, perhaps when conditions in the Near East were drier and semiarid lands inhospitable—a classic case of deserts having climatic lungs that pushed people to the margins.

Radiocarbon dates, Neanderthals, and Cro-Magnons

When did the first modern humans arrive in Europe and Eurasia? How long did the Neanderthals survive alongside them, a few thousand or tens of thousands of years? The answers to these hotly debated questions depend on an extremely refined chronology, which, whether we like it or not, is based almost entirely on radiocarbon dating. Chemist Willard Libby of the University of Chicago developed the carbon-14 (C14) method during the 1950s. He used a Geiger counter to count the decay of C 14 isotopes produced in the earth's atmosphere by the reaction of neutrons to nitrogen. When an organism dies, it contains the same proportion of radiocarbon as that in the atmosphere at the time. This disintegrates slowly; after 5,730 years, only half the original amount is left, and so on. The amount of C14 in a forty-thousand-year-old sample is minuscule, which means that the events described in this chapter lie at the very outer limits of the radiocarbon timescale. Libby assumed that the amount of C14 in the atmosphere remained constant through time, which we now know it did not. This means that all radiocarbon dates are not only statistical estimates but also nothing more than radiocarbon ages, rather than dates in calendar years. They require calibration to be converted into a historical chronology of years before the present.

Radiocarbon dating is now much more accurate than in Libby's day. For example, by using accelerator mass spectrometry (AMS), a laboratory can date tiny samples such as a single seed or a speck of carbonized wood struck in the mounting of a bone point. This makes for much more accurate dates than the handfuls of charcoal needed for dates processed thirty years ago, as such handfuls could come from a hearth used over a long time, or from an entire occupation layer. There are still, however, major issues with potential sources of contamination and conditions in the layer where the sample came from. Some of these can be resolved by very careful excavation

and recording. Other sources of contamination require precaution-
ary work in the laboratory. Many problems remain, but it's safe to
say that the science of radiocarbon dating is much more sophisticated than it was even a generation ago.

The first radiocarbon calibrations came from comparing C14
readings against tree-ring chronologies obtained from ancient European oaks and other trees. Such calibration curves provide accurate
timescales for the closing millennia of the last glaciation and all of
history since the Ice Age. The problems intensify with earlier times,
for which our knowledge of atmospheric C14 levels is still fragmentary. Researching potential calibrations bristles with highly technical
difficulties and depends heavily on highly accurate chronologies
obtained from such archives as coral growth rings, as well as ice and
deep-sea cores. Atmospheric C14 levels were extreme during the last
glaciation and especially during the period when moderns replaced
Neanderthals in Europe, further complicating the calibration issue.

In recent years, a calibration curve has emerged from comparisons
of coral sequences, combined with data from Greenland ice core
GISP-2 and the remarkably precise deep-sea core from the Cariaco
Basin, off the coast of Venezuela. The chronological links between
events recorded in these cores and other sources come from dating carried out on Greenland ice cores using uranium-thorium C14 measure-
ments. These chronologies were reasonably satisfactory and are being
refined even further by stalagmite records from the Hulu cave, in
northern China.[15] The fine-tuned calibration curve available today
provides us with at least a provisional calendar chronology for the period 45,000 to 25,000 years ago and for the events described in this and
subsequent chapters. Two useful chronological checks have come from
well-dated volcanic events: the Campanian eruption, in Italy, dated to
39,395 plus or minus 51 years ago, and an ash zone in the North Atlantic dating to 54,500 plus or minus 1,000 years before the present.

No one would describe the latest calibration curve as the final
word on the subject, but it is the very best we have. Any chronology

in calendar years is highly provisional, but we know enough to be fairly certain that some radiocarbon dates during the Neanderthal-Cro-Magnon encounter are as much as four thousand to five thousand years too young. We are fortunate to have the well-documented Campanian ignimbrite eruption as a chronological marker.

The long-term solution involves obtaining dates from different radiocarbon laboratories, as a precaution against persistent errors, and using other dating methods, like paleomagnetism and luminescence dating, in tandem with carbon-14. All we can do at the moment is express the dates in ranges of millennia, in the hope that the ranges will be narrowed dramatically in future years. As we shall see, the problem of fine-grained dating is all-important when studying potential contacts between Neanderthals and newcomers.

The latest radiocarbon calibration curve can be viewed at http://www.calpal.de.

What, then, was the origin of the Aurignacian? Aurignacian tools do occur in Near Eastern rock shelters, but somewhat later than they do in Europe, as if some groups spread southward, perhaps during the colder conditions after thirty-nine thousand years ago. Most likely, the Aurignacian cultural tradition developed in Europe itself out of the tiny earlier modern tool kits, such as we see in the levels at Kostenki 14 and elsewhere. Perhaps the changeover resulted from disruptions caused by the Campanian eruption, which may have caused hitherto isolated modern populations to move into closer juxtaposition, into circumstances where they developed a common adaptation to increasingly cold, and often variable, climatic conditions. What is striking is that the Aurignacian appears almost simultaneously right across Europe. It is also noticeable that the first settlement of the same vast region took hold extremely rapidly by prehistoric standards, during a mere three thousand to five thousand years. And, looking at the wider picture, the ancestors of the Cro-Magnons moved from northeast Africa as far as the outer recesses of western Europe in something less than ten thousand years.

WHAT, THEN, ABOUT the Neanderthals? We know from site distributions that the densest Aurignacian populations clustered in better-favored areas such as the Danube Valley and especially in southwestern France. Here the southern limits of the steppe-tundra were highly productive. The nearby Atlantic Ocean ensured less-severe winters and a longer growing season. Above all, the habitats of this region of deep river valleys and fast-flowing rivers supported a magnificent bestiary. Some river valleys, like the Dordogne and the Vézère, were major migration routes for reindeer herds and also supported numerous bison and wild horses. If stratified occupation levels are any guide, late Neanderthal populations were especially dense in these valleys, so much so that they may have survived here long after modern groups had moved into the west, perhaps for two thousand years or more.[16]

At first, it seems, the newcomers kept clear of the Neanderthals, moving into northeastern and northwestern Spain by forty-one thousand years ago, but apparently not settling in the Dordogne. That there were contacts between natives and immigrants is beyond question, even if they kept their distance—and there are the artifacts to prove it, among them a distinctive form of backed knife blade known as the Châtelperron knife and flakes with toothlike edges (figure 6.5). These artifacts were found at La Grotte des Fées, in Châtelperron, Allier, in east-central France, in an occupation layer sandwiched between Neanderthal layers, followed by an Aurignacian occupation.[17]

The Châtelperron knives and other tools were puzzling because they combined features of Neanderthal and Aurignacian tool kits. At another site where the knives were found, La Grotte du Renne, at Arcy-sur-Cure in central France, Châtelperron toolmakers even made carefully fashioned bone and ivory tools, and people wore perforated animal teeth. Such "Châtelperronian" occupation levels also occur elsewhere, mainly in central and western France, with some rare outliers in northern Spain. Then Neanderthal bones were found in the Châtelperronian

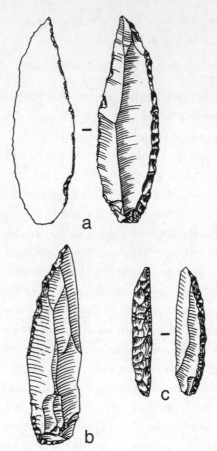

Figure 6.5 *Châtelperron knives, which may have been stand-alone tools or artifacts with blunted backs mounted on spear shafts. After G. Henri-Martin, 1957.* La Grotte de Fontéchevade. *First part. (Paris: Masson: Archives de l'Institut de Paléontologie Humaine, Memoire 28).*

levels at Arcy-sur-Cure and another site, Saint-Césaire, in western France. Were, then, these seemingly Aurignacian types of artifacts made by Neanderthals or received in exchanges with adjacent Aurignacian groups?

Radiocarbon dating is precise enough that we know that the Châtelperronian levels at seven sites in France and northern Spain date to about forty-one thousand years ago. Early-modern occupations in northern Spain are in the same time frame. Not only that, but the

Neanderthal levels underlying the Châtelperron horizon at Saint-Césaire date to about forty-three thousand years ago, a date that coincides with that of an equivalent level at the type site, Le Moustier itself. Furthermore, three or four sites in France and northern Spain show that people making Châtelperron tools and people occupying early Aurignacian levels actually visited the same caves at different times; so frequently that their artifacts alternate in the stratified occupations.

Perhaps just as significant, radiocarbon dates also tell us that people making Aurignacian artifacts only settled in the relatively densely settled area of southwestern France occupied by the Neanderthals after thirty nine thousand-years ago, later than they settled in Spain. The seeming reluctance of Aurignacian groups to live in the region may well be connected to climatic and environmental factors, but above all to the exceptionally high density of Neanderthal populations in the area. It is precisely here that we find the highest concentrations of Châtelperron tools.

Southwest France was one of the areas where Neanderthals survived the longest, in a favored, relatively sheltered enclave, where hunting territories were comparatively small and the age-old rhythm of their lives continued much as it always had. On the other side of the Pyrenees, the moderns lived in different, slightly warmer and drier environments, rarely penetrating north of the mountains. There were occasions when they did, perhaps as a result of short-term climatic shifts and population growth, or because of the constant ebb and flow of hunter-gatherer life. It was then that there were sporadic contacts with Neanderthals.

The moment *Homo sapiens* spread across Europe, the Neanderthals were at a disadvantage. Premoderns and moderns hunted much the same quarry, even if the latter's weaponry was lighter and had greater penetrating power. Neanderthals relied on ambushes; so did the moderns. Nevertheless, the main differences between the two were insurmountable. Aurignacians had fully articulate speech and all the cognitive abilities of *Homo sapiens*. The Neanderthals were no slouches, but their communication skills were more limited. Above all, they lacked the qualities of imagination and self-awareness found in moderns, the latter's ability to plan far ahead, and all the complex social mechanisms found in Cro-Magnon society. The Aurignacians lived in progressively larger groups, with enhanced social interaction and a longer life expectancy. Innovation was commonplace; knowledge passed effortlessly from one longer-lived generation to the next.

To the Cro-Magnons, the Neanderthals were probably little more than a lesser form of themselves, something considered inferior. The former may even have been repelled by the latter's hirsute, squat appearance. The Cro-Magnons may also have feared the Neanderthals' strength, especially at close quarters, as a Neanderthal would certainly have prevailed in a hand-to-hand fight to the death. However, the Neanderthals probably feared the razor-sharp bone-tipped spears of the Aurignacians, which could deliver a fatal wound from far beyond the range of a throwing stick or a heavy stone-tipped spear.

Each probably observed the other from a distance, cautiously and silently, perhaps with unspoken understanding between them. Quietly, the Neanderthals would have observed the more-refined technology of their neighbors, sometimes picking up abandoned tools from rock shelter floors. Judging from modern-day controlled experiments at replicating their artifacts, the Neanderthals were competent artisans who could "read" stone as easily as you and I peruse a book, a skill learned by today's stoneworkers. So it was easy for them to copy simple tools like bone points and backed knives and incorporate them into their daily tool kits. But their qualities of innovation, of original thought, never allowed them to elaborate the newcomers' technology and create tools and weapons that competed with Aurignacian weaponry.

Extinction was a slow death, no dramatic event, that unfolded on the edge of the Aurignacian world.[18] Some bands may have perished in angry confrontations over hunting territories, where men died and the Neanderthals slowly gave way, retreating to the margins of their former lands or occupying hitherto uninhabited landscapes. At Les Rois cave, in southwest France, a Neanderthal jawbone displays the kind of butchery marks made with stone tools that resulted from the removal of the tongue and teeth. Archaeologist Fernando Rozzi believes that a Neanderthal met a violent end at the hands of modern humans, who then removed the teeth and perhaps the skull and jaw as trophies. This fascinating discovery does not necessarily mean that the Cro-Magnons regularly ate Neanderthals, if at all, but it certainly hints that competition from modern humans contributed to the extinction of their forebears.[19]

Other Neanderthal groups must have simply faded into the countryside and eked out a living as far away from the newcomers as they could, based in remote caves, isolated valleys, and locations where game was less abundant. As numbers fell and men died in hunting accidents, and bands merged with one another, Neanderthal populations must have become more and more isolated, to the point where all the childbearing women died and band after band slowly vanished. There was no one dramatic moment of extinction but numerous such events, all of them different, all of them involving unique circumstances. In the end, there were small pockets of Neanderthals, carrying on in ever-diminishing numbers, confronting increasingly severe climatic conditions. During their heyday, they could adapt comfortably to extreme cold, but a combination of falling numbers in always small, often isolated groups and competition with more-numerous modern humans may have made the ultimate difference and led to extinction.

We will never know where the last Neanderthals lived, because they were so widely dispersed over enormous distances. Spain was one of the last places where they survived, and we can glimpse them on the verge of extinction. Gorham's Cave, at the foot of the Rock of Gibraltar, was home to Neanderthal groups from 125,000 years ago until about 30,000 years before the present—the final date is uncertain.[20] The cave then lay on the edge of a coastal plain covered with scrub savanna, a place where a hunting band could prey on small animals, sea mammals, and fish. After 30,000 years ago, intensifying cold turned the savanna into semiarid steppe, and a diminishing Neanderthal population vanished. That they survived so long in places like Gorham's Cave is a tribute to their ingenuity and opportunism, but, like so many other mammals, they finally became extinct, leaving the Cro-Magnons as the sole masters of Europe.

CHAPTER 7

The Realm of the Lion Man

FRANCE, SUMMER, THIRTY-SEVEN THOUSAND years ago. The men have traveled far, winding cautiously through deep river valleys and across fast-flowing streams. They keep a sharp eye out, spears at the ready, for lions hiding in dark clearings among the trees that press on the water's edge. A thin plume of wood smoke rises among the trees from a small rock shelter at the foot of an outcrop. As the men approach, the *click*, *click* of stone against stone echoes against the dark boulders. Cautious greetings are exchanged, dried meat is offered, a meal prepared. As night falls, the women stoke the fire and the bargaining begins. The visitors produce some perforated seashells carefully wrapped in leather pouches. Their hosts hold the mollusks up to the firelight, twisting and turning them in their hands. A boy vanishes into the shadows, then returns with a handful of carefully shaped flint lumps. Now it's the turn of the visitors to examine them carefully, massaging the stone, feeling for invisible defects in the rock. The bargaining continues for hours, interspersed with gossip, laughter, and the inevitable hunting stories. The piles of stones and seashells grow slowly. The next morning, the men depart, their pouches laden with fresh toolmaking stone . . .

From the moment that the Cro-Magnons set foot in their new homeland, ways of doing business changed fundamentally, partly because of fluent speech and also because of superior intelligence, great mobility, and personal connections that extended far over the horizon. We know this because, unlike the Neanderthals, the Aurignacians acquired their toolmaking stone from afar, often from fifty miles (80 kilometers) or more away. They had to because of their technology. Fine-grained rock is near essential for smaller blade blanks and is more easily flaked into cores that yield numerous blanks. Think of even a single core as a stone savings bank, carried, perhaps, in a leather bag at the hip, then taken out and used whenever a projectile point broke or a woodworking tool was needed at short notice. A hunter would break a spear point, try to repair it, then throw it away and strike off another blank to fashion another one in a few minutes.

Artifact sourcing tells us that the Aurignacians traveled regularly to convenient outcrops, perhaps visiting the same location at specific times of year, or just when they were nearby. Perhaps, too, the outcrop might lie outside their hunting territory and be the property, as it were, of another band. Now ties of kin came into play, or one-to-one relationships developed between the individuals who quarried the stone and those who needed it. We know that perforated teeth, seashells, and other ornaments traveled over long distances before Aurignacian times, so it's easy to imagine that these one-to-one contacts involved the exchange of gifts and the forging of long-term relationships that passed from one generation to the next. Over the centuries and millennia, inconspicuous exchange networks linked band to band, territory to territory, over enormous areas of the Cro-Magnon world.

Fine-grained rocks were probably far more than just artifact material. If the Cro-Magnons were anything like the Australian Aborigines, admittedly a far-fetched analogy but perhaps a valid one, major outcrops of toolmaking stone may have been places of power, the stone itself a powerful medium. The Yoingu Aborigines of eastern Arnhem Land, in Australia, are the traditional owners of the quartzite outcrops at Ngilipitji. They believe that the stone grows underground, where it becomes pregnant with small stones, like eggs. The quartzite sparkles, a quality that it shares with blood, which also has a brilliance known as *bir'yun*. Quartzite may make efficient spear points, but its aesthetic value far exceeds any practical use. These outcrops are far more than just rock: because of their magical properties they linked people in symbolic networks. The same must have been true in late Ice Age times.[1]

The Australians also value natural pigments such as ocher, a form of iron oxide that comes in hues from yellow through red and brown. Aurignacians also placed a high premium on ocher; it abounds in their settlements. They ground the ocher into powder, then mixed it with water or some other liquid, such as spittle or blood. The resulting pigment had numerous uses, many of them of social or symbolic significance: as body paint, for decorating artifacts or coloring one's hair, and also for cave paintings. Again, if the Australian Aborigines are any guide, ocher sources were of great importance, for the Aborigines consider the pigment to possess curing properties and to strengthen its user. Whether the Cro-Magnons associated these same properties with it is, of course, a matter of guesswork. The Aborigines live in arid landscapes and believe that the deep red of ocher is the blood of their ancestors. Red ocher was in common use in the ancient Near East. Could it have had the same symbolic association with ancestors as it does among Australians, as many Aurignacian and later burials were sprinkled with red ocher?

If toolmaking stone had other properties, was considered a living thing, and was passed from hand to hand over long distances, then the Aurignacians' society was very different from that of their predecessors.

DIFFERENT INDEED, AS even a cursory glance at the *Löwenmensch* reveals (see color plate 1). A man with a lion head appears to straddle the human and animal realms, as if he has the ability to transform himself from a hunter into the hunted. There is something mystical here, a connection between the living and the supernatural quite unlike anything we have encountered before. The Lion Man is the oldest known example of an imaginary being.

Löwenmensch comes from the Hohlenstein-Stadel (hollow rock barn) cave, which lies in the cliff of that name at the southern edge of the Lone Valley, in the Swabian Alps of southern Germany. In 1939, archaeologists Otto Völzing and Robert Wetzel discovered an Aurignacian settlement in the cave, which was remarkable for its numerous fragments of mammoth ivory. The outbreak of World War II abruptly

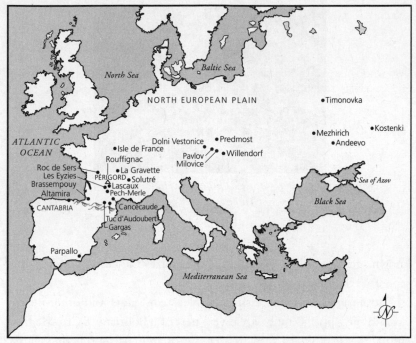

Figure 7.1 *Map showing sites mentioned in chapters 7 to 11.*

ended the excavations, which did not resume until 1954, when they lasted for another seven years. The Aurignacian layers proved to be more than thirty-four thousand years old.[2]

The Hohlenstein collections lay virtually forgotten in the Ulm Museum until the late 1960s, when Joachim Hahn began piecing together some larger ivory fragments, which had come from a niche in the cave. To his astonishment, they formed a human figure with a lion's head. When Elisabeth Schmid and Ute Wolf finally completed the reconstruction in 1988, they had pieced together an elongate carving just over 11 inches (28.1 centimeters) long. The left arm bears slanting marks that may represent scarring or tattoos. The upright posture is definitely human, but the feet seem pawlike. They do not allow the figure to stand on its own, as if it was meant to be placed in a hole or leaned against something. Almost certainly, the figure is male, for there are no breasts. The Lion Man is both an animal and a human at the same time, the earliest

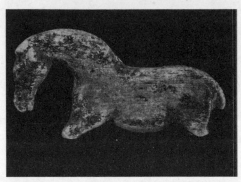

Figure 7.2 *Vogelherd horse. Photo by Hilde Jensen. Institut für Urgeschichte, Universität Tubingen.*

known such figure in the world. He epitomizes the fluid boundary that separated the Aurignacians from their prey.

The Aurignacians who visited Hohlenstein-Stadel and other nearby caves in the Upper Danube area were part of a flourishing Cro-Magnon culture that exploited a very broad range of animals, among them large, formidable beasts such as aurochs, bears, mammoths, and rhinoceroses. They used mammoth ivory to make pendants, rods, and other artifacts, as well as ivory beads, all unique to this region. They also carved small animal figurines, which represented only a small portion of the bestiary around them—almost invariably large, powerful animals. The carvings always show obvious facial features such as the mouth, ears, and eyes, sometimes even a distinctive expression. The body is well represented, but the extremities are reduced or omitted altogether. Some of the carvings bear geometric patterns scratched on their flanks: dots, curved lines, angles, and other motifs. Some may have served as pendants or been carried in leather pouches. Modern experiments have shown that it would have taken about forty hours to fashion a horse figure found in the Vogelherd cave (figure 7.2).[3] The Hohle Fels cave, also in southern Germany, has yielded a female figurine in mammoth ivory 2.4 inches (6 centimeters) tall. The woman has massive breasts, an enlarged and explicit vulva, and large belly and thighs, very much in the tradition of later female figurines described in chapter 9. Clearly, the

Plate 1: The "Lion Man" figurine from Hohlenstein-Stadel, Baden–Württemberg, Germany. Ht: 296 mm. THOMAS STEPHAN, COPYRIGHT © ULMER MUSEUM.

Plate 2: Grotte de Chauvet: panel with horses and woolly rhinoceros. MINISTÈRE DE LA CULTURE ET DE LA COMMUNICATION. PRÉFECTURE DE LA RÉGION RHÔNE–ALPES. DIRECTION RÉGIONALE DES AFFAIRS CULTURELLES.

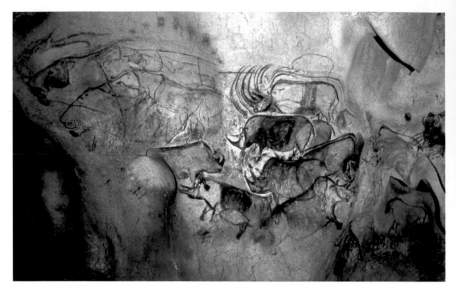

Plate 3: Grotte de Chauvet: frieze of rhinoceroses. MINISTÈRE DE LA CULTURE ET DE LA COMMUNICATION. PRÉFECTURE DE LA RÉGION RHÔNE-ALPES. DIRECTION RÉGIONALE DES AFFAIRS CULTURELLES.

Plate 4: Grotte de Chauvet: lionesses. COURTESY: MINISTÈRE DE LA CULTURE ET DE LA COMMUNICATION. PRÉFECTURE DE LA RÉGION RHÔNE-ALPES. DIRECTION RÉGIONALE DES AFFAIRS CULTURELLES.

Plate 5: Reconstruction of Gravettian occupation at the Abri Pataud, France, c. 24,000 years ago. ILLUSTRATION ERIC GUERRIER. MUSÉE DE L'ABRI PATAUD AUX EYZIES-DE TAYAC, DORDOGNE AND INSTITUT DE PALÉONTOLOGIE HUMAINE. DRAWING EXECUTED UNDER THE SUPERVISION OF PROFESSOR HENRY DE LUMLEY.

Plate 6: Lascaux Cave, France. An aurochs leaps over a frieze of small horses. THE BRIDGEMAN ART LIBRARY.

Plate 7: Lascaux: two bison. Martin Jenkinson/Alamy.

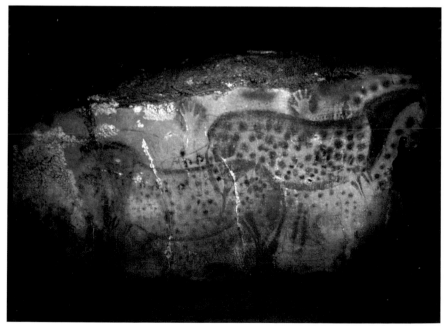

Plate 8: Pech Merle: spotted horses. Centre de préhistoire du Pech Merle, Cabrerets, France.

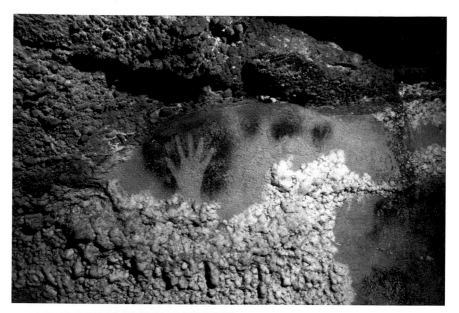

Plate 9: A hand impression from Pech Merle, formed by blowing red ocher onto the cave wall. BILDARCHIV PREUSSISCHER KULTURBESITZ/ART RESOURCE, NY.

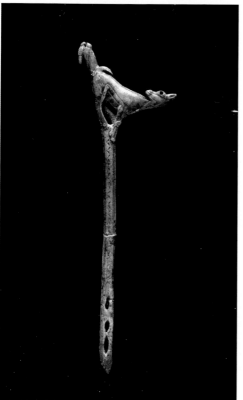

Plate 10: Spearthrower of an ibex from Mas d'Azil. RÉUNION DES MUSÉES NATIONAUX/ART RESOURCE, NY.

Plate 11: Clay bison at Tuc d'Audoubert. SISSE BRIMBERG & COTTON COULSON, KEENPRESS/ NATIONAL GEOGRAPHIC IMAGE COLLECTION.

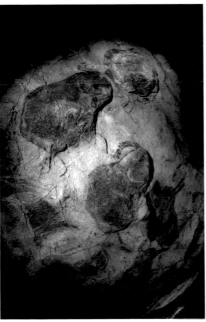

Plate 12: The painted chamber at Altamira, Spain. SISSE BRIMBERG & COTTON COULSON, KEENPRESS/NATIONAL GEOGRAPHIC IMAGE COLLECTION.

Plate 13: Polychrome Altamira bison. ROBERT HARDING PICTURE LIBRARY/ALAMY.

Plate 14: An artist's impression of a Cro-Magnon reindeer hunt. COPYRIGHT © GROUPE FLEURUS. ILLUSTRATIONS DE CHRISTIAN JEGOU, LASCAUX.

Plate 15: An artist's impression of a Cro-Magnon summer camp. Copyright © Groupe Fleurus. Illustrations de Christian Jegou, Lascaux.

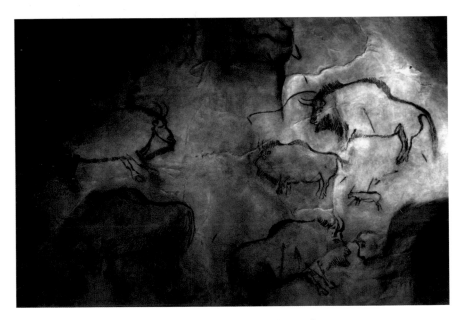

Plate 16: Bison frieze at Niaux. Charles Jean Marc/Corbis Sygma.

practice of sculpting women with exaggerated sexual features was part of Aurignacian art from very early times.[4]

The Lion Man and these other figurines are some of the earliest art in history. His serene animal-human countenance and the well-worn models of dangerous prey set the Aurignacians apart from the Neanderthals. Just the presence of a figure that is part human, part animal adds an entirely new dimension to our perceptions of the Cro-Magnons. These are far more than mere art objects. They are symbols of a burgeoning and obviously complex range of spiritual beliefs.

I've cast the Neanderthals as the quiet people, silent observers, soon in the minority, often content to watch from a distance. They were probably unobtrusive neighbors who had little to say to the moderns. Neanderthals had some form of speech, but whether this was sufficient for them to communicate on a day-to-day basis with their modern competitors is unknown. Did they, for example, speak some form of "pidgin Aurignacian," sufficient for them to improve their linguistic skills in a limited way?[5] Even if they did, communication was probably limited and confined, for the most part, to gestures and simple words. We know that in several areas of Europe, especially southwestern France, Neanderthal artisans copied such tools as Châtelperron knives and perhaps even adopted some different hunting practices. At Pech de l'Azé cave, near Eyzies, Neanderthals may have used black pigment for body decoration, but that does not mean that their minds were able to grasp the complex reasoning and intellectual world of their new neighbors. Nor did they necessarily consider personal adornment a form of social expression.

Neanderthals used red ocher and perhaps body paint, but we don't know if they learned of them from the Aurignacians. They may have acquired some perforated teeth by exchange, or even quietly stolen them. Both forms of adornment do not place any great intellectual demands on the user, unless they are used within a social context. Under the latter circumstance, the surface of the body becomes a form of language for expressing social identity as it changes through life—and at death.

There are also ritual forms of adornment used only occasionally, decoration worn habitually, and symbols that are associated with hunting or childbirth. The figurines and carvings made by the Aurignacians of Vogelherd and other Swabian caves were fashioned of bone and ivory from carnivores such as lions, mammoths, and wolves. This may be no coincidence—that the most dangerous predator of all would use raw material from other formidable beasts for sculptures that would be on social display. Such nuances would have been beyond the mental imagery of the Neanderthals.

The Neanderthals may have copied the bodily adornments of their neighbors, but they probably did nothing more than pretend they were Aurignacians. Theirs was selective borrowing, which ignored images, elaborate burials, and other complexities of modern human life. The Neanderthals possessed an entirely different form of consciousness. Their relationships were based on age, sex, experience, and strength. Much of their imagery was linked to basic motor skills driven by very limited imagination. Cro-Magnon imaginations allowed them to translate mental images into figures engraved or painted on a wall, or carved out of a piece of ivory. Their imaginations ranged and soared, found expression not only in motor skills but also in all kinds of other things: the shifting three-dimensional movements of clouds against a blue sky, the colors of reindeer at different seasons, the subtle movements of shadows from a blazing hearth flickering against a rock shelter wall. The Aurignacians conjured up a supernatural realm of ancestors, spirit beings, mythical animals, movement, shifting lights, and brilliant colors. Everyone, man or woman, young or old, adult or child, possessed a lively imagination and, of course, used it in different ways.

The Cro-Magnons' sense of heightened consciousness not only conceived of a supernatural realm but also prepared them for it. The forces of the natural world were always around them, sometimes benign, often hostile, threatening, cajoling, and offering precedent and encouragement. One passed through life navigating these hazards and benefits using skills and philosophies that were handed down from one generation to the next by word of mouth, through storytelling, chant, and ritual.

The dramatic find of four bone-and-ivory flutes from the Hohle Fells cave dating to 35,000 years ago testify to their musical abilities. One of the flutes was made of griffon vulture bone, an instrument 8.6 inches (21.8 centimeters) long with five holes and a notched end. Three others, made of mammoth ivory, may have produced deeper sounds but were much harder to manufacture. The process involved splitting the ivory, hollowing it out, making the holes, then fitting the two halves together with notches and some form of adhesive.

As these flutes testify, no longer were humans just predators in the food chain. They were now dynamic partners in a world peopled by animals they considered to be living beings to be treated with respect, prey that fed them but also provided important powers in a world where the living and the supernatural were as one.

I believe this was the most fundamental difference between the Cro-Magnons and the Neanderthals, and one that probably caused them to perceive their neighbors as inferior, probably as little more than animals. The Cro-Magnons' ability to conjure up mental images and manipulate them must have colored their relationships with premoderns and may have given them a sense of superiority. The Neanderthals had survived more than 160,000 years in harsh, ever-changing environments, sometimes of extreme cold. Their society changed slowly and in limited ways, reflected in the development of some regional tool kits whose artifacts nonetheless never strayed far from the basics of their technology. Their challenges were, above all, environmental, for they had no human competition until the Aurignacians came along. The newcomers were efficient, highly mobile, and living in an entirely different form of society, which boasted far more intricate social relationships not only with fellow band members but also with others living some distance away. Neanderthals could never imitate the intangibles of Cro-Magnon society, the complex dynamics that now governed relationships between individuals and groups. This new, ever-changing partnership between the living and the supernatural holds the clue to the Lion Man, depicted as a part-human, part-beast mythic creature.

CZECH REPUBLIC, MIDWINTER, THIRTY-FIVE THOUSAND years ago. Huddled around a blazing hearth under the rock shelter's overhang, the band is swathed in furs inside the hide tent pitched against the back wall. Children snuggle up against their parents. Outside, the temperature is well below zero. Inside, it's just above freezing, but smoke hovers near the smoke hole high overhead. Most of the small band are young, the oldest in their early twenties. One man stands out, his weathered face visible under his parka hood. He is older, one arm stiff from a hunting accident long ago, and his expression conveys the wisdom of years. He is a storyteller, a mentor for the band. He tells of the ancient day when the world came into being, created by intelligent commonplace animals, their prey. In time, these animals fashioned not only rocks and valleys, rivers and forests, but also human beings, men and women. The elder speaks for hours, his voice rising and falling as he pauses for emphasis, gestures dramatically, or describes the character of a human or an animal. Everyone has heard the story before, but his audience is as attentive as it was the first time. The story prompts the imagination, stresses ancient skills and the spiritual relationships that nurture the hunters and their world. As he speaks, they paint the cosmos, the imaginary world of his tale, in their minds. And one day, when the elder has become a revered ancestor, one of his descendants will tell the same tale. Neither daily life nor the story will stay completely the same, but there will always be a close, and constantly changing, relationship between the two . . .

Who, then, was the Lion Man? Was he a mythic being or a living person with unusual supernatural powers? The lion-headed human figurine implies some form of transformation, an effortless passage between the living and supernatural realms. We will never, of course, be able to decipher the intended purpose of the Lion Man, but there is a good chance that he represented a "person of power" in his society, someone who maintained good relationships between the living and the forces of the supernatural world.

People of power flourish in numerous human societies to this day. They are often generically, and misleadingly, called shamans, a term that comes from the Siberian herder word *saman*.[6] Siberian *samans*

negotiated between the world of living people and other realms of existence, such as those inhabited by deities, ancestors, and the spirits that
controlled nature. They passed between the two realms in trances in-
duced by self-hypnotism or hallucinogens and were well aware of the
power of vivid performance and imagined fear. The shamans often used
elaborate paraphernalia: exotic headgear, feathers and animal fur,
masks, amulets and pendants that hung from their clothing, and above
all rattles and drums. In short, they were people of power.

Not all people of power can be called shamans, but all such individuals share some characteristic behaviors. They act out rituals of transformation, of passage from one realm to another, in societies where
boundaries between humans and their environment, between people and
their ancestors and the realm of the supernatural, do not exist. This per-
meable continuum between the living and the supernatural is common to
many hunter-gatherer and agricultural societies today and has been
commonplace for an immensely long time. The mere presence of the
Lion Man in the Hohlenstein makes it likely that such a continuum was
already in existence in Aurignacian society. This does not, of course,
mean that there were shamans in Cro-Magnon society. But there must
have been some individuals who possessed compelling relationships with
the supernatural, which was an integral part of Cro-Magnon existence.

We should not be surprised by this, for Aurignacian life, like that of
both Neanderthals and later Cro-Magnon societies, revolved around
the realities of the passing seasons, the seasons of plant foods, and the
unfolding litany of birth, life, and death. Every hunter-gatherer society
in history lived a basically similar cyclical routine, which repeated itself
endlessly. Winters might become longer and colder, reindeer herds
move, or nuts become unusually plentiful, but the rhythm of life continued unchanged, just as it had for one's forebears and would for generations as yet unborn. People thought of themselves as part of a living
world, where animals, plants, and even landmarks and inanimate objects had lives of their own. The environment lived and surrounded one,
defined by tangible and intangible forces or personalities, whether human or not. To live in such a way required powerful imagination, the
ability to conceptualize, to chant, to make music, and to tell tales that

validated human existence and explained the natural order of things. As part of this, it was inevitable that some individuals acquired, or were perceived to possess, unusual supernatural powers, which involved communication with the world of the intangible. We know that today such people acquire their powers through solitary quests, visions, or compelling experiences. There's no reason to believe that Cro-Magnon societies didn't nurture their own people of power.

People of power have always been men and women of charisma and unusual personal qualities, often feared for their powers. It's a gripping experience to encounter one in the fullness of his or her powers. Many years ago, I watched a spirit medium among the Dande people of central Africa's Zambezi Valley enter a trance and deliver the genealogy and history of their ancestors in an accurate recitation that convinced the assembled crowd that the ancestors indeed spoke through him. His was a precarious role, for the ancestors had the welfare of the people at heart before anything else, and he had to maintain a reputation for credibility and authenticity. This particular medium had a flourishing practice for many years, its success based partly on his shrewd assessment of public opinion. Sometimes, people of power have been curers and magicians, and they may have had this role in Cro-Magnon society, fully capable of bringing misfortune or success. Judging from rock paintings in remote chambers far from daylight, some of them may have quested in solitary darkness far from the living world.[7] Invariably, such individuals have been performers, well aware of the power of the drum and the rattle, of flickering light and dark shadows, of storytelling and dance. They listened to public opinion, issued pronouncements, and performed hunting rituals. Above all, they were the repositories of huge repertoires of oral tradition and experience, of legend and correct ritual observance. Whatever name they are given, individuals with supernatural powers and the ability to transcend the human realm were catalysts in all kinds of hunter-gatherer societies. The Lion Man is a strong hint that at least some transformation rituals unfolded among the Aurignacians of the Upper Danube. These rituals would have involved a close relationship between the humans and powerful animals that was fundamentally different from anything experienced by the Neanderthals. All that we know about the

Cro-Magnons' rituals and beliefs comes from their art. They created sentences and narratives, with infinite possibilities and novelties that changed by the day. Their spiritual life possessed open-ended sophistication, much of it passed from one generation to another, probably by repetitious mnemonics.

FOR YEARS, THE experts assumed that the Aurignacians were tentative artists, perhaps limited in their skills or somewhat backward in their ceremonial life. There were few cave paintings that could be attributed to them, and they did not engrave antler or bone with any sophistication. The discovery of the Lion Man came as somewhat of a shock to this long-established theory. Then, on December 18, 1994, three speleologists with an interest in archaeology, Eliette Brunel Deschamps, Jean-Marie Chauvet, and Christian Hillaire, crawled into a small opening in the Cirque d'Estre gorge, in southeastern France's Ardèche. The entrance was a mere thirty-one inches (eighty centimeters) high and twelve inches (thirty centimeters) wide, but it led to a narrow vestibule with a sloping floor. The three explorers felt a draft flowing from a blocked duct. They pulled out the boulders that were blocking it and saw a vast chamber nine feet (three meters) below them. After returning with a rope ladder, they descended into a network of chambers adorned with superb calcite columns. Calcified cave bear bones and teeth lay on the floor near shallow depressions where the beasts had hibernated. Suddenly, Deschamps cried out in surprise. Her lamp shone on two lines of red ocher, then a small mammoth figure.

The group soon penetrated into the main chamber and came upon more paintings: hand imprints and figures of mammoths and cave lions, one with a circle of dots emerging from its muzzle. As they gazed at the paintings, the three explorers were "seized by a strange feeling. Everything was so beautiful, so fresh, almost too much so. Time was abolished, as if the tens of thousands of years that separated us from the producers of these paintings no longer existed."[8]

Grotte de Chauvet had lain undisturbed since the late Ice Age. The original entrance is blocked, but near it appear three red paintings of cave bears, identified by their steep foreheads. A natural bulge in the rock forms the shoulder of the largest bear. Chauvet had been a bear cave long before being visited by Cro-Magnons. Carefully arranged bear bones and skulls hinted that the human visitors had paid homage to them. The cave had been abandoned after thousands of years of use, perhaps when the entrance collapsed unexpectedly, leaving the floor intact and undisturbed. Hearths on the floor looked as if they had been extinguished the day before. Flaming torches had been rubbed against the wall to remove the charcoal so they would flare anew. A little farther on in the chamber lay a slab that had fallen from the ceiling. A bear skull had been set on top of it; the remains of a small fire lay behind it. More than thirty calcite-covered and intentionally placed bear skulls surrounded the slab. In an end chamber, the three explorers came across a thirty-foot (ten-meter) frieze of black figures dominated by lionesses or lions without manes, rhinoceroses, bison, and mammoths. Experts were called in and radiocarbon samples were taken from the paintings. To everyone's astonishment, the earliest paintings were about thirty-three thousand years old. New calibrations of radiocarbon dates for the cave paintings run to thirty-six thousand years ago. If these dates are correct, it would have been the Aurignacians who visited and painted Grotte de Chauvet. (For illustrations of Chauvet paintings, see plates 2 to 4.)

People left their imprints on the walls. In a side chamber fairly close to the entrance, an Aurignacian visitor left clusters of red dots on a prominent rock face. He or she made the dots by covering the palm of his or her right hand with red paint, then applying it against the wall. Sometimes you can even discern traces of the fingers. Almost all the dots are at the same height, as if they were made by one individual. A shorter person, perhaps an adolescent, made another panel. In another place, a large slanting panel extends over nearly forty feet (twelve meters). Here, an artist painted red figures of rhinoceroses. A long wavy line extends over some handprints, made by covering the hand with paint and pressing it against the rock. There are negative handprints, too, impressions depicting only the contours of the hand, the inside being left blank. The entire panel is an intricate palimpsest of animals, geometric signs, and handprints, all executed in red.

The Chauvet enigma

Grotte de Chauvet is among the most spectacular of all Cro-Magnon painted caves. The artists who painted its walls had a sophisticated understanding of perspective and other artistic principles, so much so that the experts first assumed that the paintings were executed less than eighteen thousand years ago, during what was then thought to be the heyday of Cro-Magnon art. The radiocarbon dates, obtained from tiny samples from a few paintings, as well as from charcoal from floor features, caused a sensation, for they pushed back use of the cave to a period between about thirty-six thousand and more than twenty-four thousand years ago. If the dates were correct, many of the paintings were the work of Aurignacian painters. After an initial gulp of surprise, most authorities accepted the art as Aurignacian. But are they correct?[9]

Dating cave art is never easy, even if one can obtain minute AMS radiocarbon samples from individual paintings without damaging them. Immediate questions arise about the Chauvet dates and what they mean. Were the paintings all executed simultaneously? Were they retouched later, the sample coming from an undetectable freshening of an image? How do we know that a cave was not used repeatedly over many thousands of years, often with long periods of disuse between visits? In the case of Chauvet, visits and painting apparently unfolded over ten thousand years or so—if the current chronology is reliable. Quite apart from very difficult issues of potential, and invisible, contamination of samples, either inadvertently in the laboratory or before they were taken from a painting, you really need very large numbers of samples from multiple images to secure even a preliminary chronology. So far, only six samples from three images date the Chauvet paintings, all of them processed by the same radiocarbon laboratory. Judging from experiences with other date sequences, Cro-Magnon and otherwise, this is probably a mistake.

Many of the Chauvet radiocarbon dates come from charcoal deposits on the cave floor, which cannot be linked directly to the paintings on the walls. They could have been used or dropped centuries, even thousands of years, before or after the painting. Very few stone artifacts come from the floor. Those that do are little more than flakes or blades scattered over the surface with no datable context. They certainly do not cluster around any individual painting or painted panel.

There are other issues as well. Rock-art experts point out that many of the stylistic and technical features of the Chauvet art are unknown elsewhere before much later times. Nor is all the art from the cave homogeneous, there being several styles, although, of course, that does not necessarily mean they were painted in later times. Other known examples of Aurignacian art on cave walls are rare, although there are cases of painted slabs found in cave deposits. How, then, does one explain the sudden appearance of sophisticated cave paintings at just one location? Certainly, painting the Chauvet art was well within the intellectual capabilities of people who had exactly the same cognitive abilities as painters of later millennia. There's another problem, though. Why are there virtually no Aurignacian sites near Chauvet? The cave is an isolated masterpiece. You can argue, of course, that people traveled to the Grotte from long distances away, or that visitors stayed in temporary camps nearby that have vanished completely, but the sheer lack of data compounds the mystery.

Where does this leave us? No one believes that the cave paintings were the work of an individual or individuals who painted everything in one orgy of creativity. People came into Chauvet many times to visit, perhaps to perform ceremonies, and to paint. Haunted as we are by the inaccuracy of radiocarbon dates, it's probably safest to say that Aurignacians were among the painters at the cave, but there may have been others later on. For the purposes of this book, I have boldly attributed many of the Grotte de Chauvet paintings to Aurignacian artists. In theory, at any rate, they were certainly capable of executing paintings as magnificent as those of their successors. The problem is to prove it beyond any reasonable doubt—and that will require a great deal more careful archaeology and radiocarbon dating.

A chamber without paintings then intervenes, leading to another section of the cave, where the paintings are in black and engraving is commonplace. Large, parallel hanging rocks bear engravings, one of them of a long-eared owl with its head turned toward the viewer while the front of its body faces the other way. This is the earliest known painting of an owl's unique ability to turn its head through 180 degrees, a telling indication of how well the artists knew their subjects. In many societies, owls have associations with supernatural powers, signal the underworld, or serve to represent human spirits after death. None other than William Shakespeare wrote of "the owl, night's herald."[10] He knew that many people called the bird the "fatal bellman" that heralded our final, deepest sleep. Among some southeast Alaskan native groups, such as the Yup'ik, owls were often helping spirits, honored in ceremonies that featured songs and masked dances that commemorated animals as living people, both dangerous and helpful.

In the end chamber of the cave, nine lions and a single reindeer face to the left. On the right, a group of seventeen rhinoceroses cavort in such a way that it's clear the artist intended to depict them as a group. All of them have small ears, arched jawlines, ball-like feet, and scraped outlines. A complete rhino stands at the top of the panel. But the artist then added two horns to the front one and three to the rear horn, each decreasing slightly in size. With a few lines, he or she created an effect of perspective, suggesting a herd of animals standing side by side, an illusion reinforced by the straight lines of their backs.

The standard of artistry is breathtaking. In a small recess, four wild horses face one another, each painted with a thick, black mane and a well-detailed mouth. The painter used his or her fingers to spread a charcoal paste that strengthens the outlines of the figures and gives them a sense of relief. In another frieze, two rhinoceroses fight with butting heads. Their horns are crossed, the movement of the struggling beasts depicted by their legs, which are only roughly sketched. Above them, the artists scraped off some earlier engravings, then drew two rhinoceroses and a stag, two mammoths, then several aurochs. Finally, a single artist painted four horses, starting with one on the left, then working to the right. He or she made clever use of the relief of the rock surface, mixing clay from the surface and charcoal to produce different hues. The edges of the figures were scraped to outline them in white. There are about forty horses in Chauvet, but other animals—seventy-six mammoths, seventy-five felines, mainly lions, and sixty-five rhinoceroses—outnumber them.

Elsewhere, a pride of sixteen lions pursues seven bison across a wall to the right of the cave's central recess. Both lions and lionesses are tense, engaged in the hunt. Only their heads and forequarters appear, but in relief, which makes them stand out from the wall. This unique scene was probably the work of a single artist and is without parallel in any later Cro-Magnon painted cave.

Almost no humans appear on Chauvet's walls, for it is powerful and often dangerous animals that preside over the cave. A few feet from the lion panel at the end of the cave, and opposite it, a hanging rock bears two black felines and a horse on one side and on the other a black creature that stands upright. The top of its body is a bison's, the bottom that of a human, with the two legs clearly visible. When you arrive in the chamber, this is the image you see first. A filled triangle is painted in front of the body, the concave base emphasized with a mark.

Another hanging rock formation extends downward 4.2 feet (1.3 meters) toward the ground. Black drawings and engravings cover its four sides, among them figures of four lions, a horse, two mammoths, and an aurochs. Then there's a strange depiction of a woman's lower body drawn in black on the lower portion of the rock. You can see her tapering legs, her pubic triangle, and her vulva, delineated by a vertical engraved mark. A black bison with a horned head stands to the right and slightly above her, the legs not ending in the usual stumps but as thin lines. The doyen of French rock art, Jean Clottes, wonders if the lines depict not animal legs but human hands, the woman and the bison representing a composite creature, part human, part bison. Perhaps, he remarks, "this is reminiscent of ancient myths, in which a mortal woman may entertain a relationship with a god or a supernatural spirit in animal form."[11]

WHAT DOES THE art mean? Controversy has surrounded Cro-Magnon rock art ever since its discovery during the late nineteenth century. Early researchers considered the animals to be art executed "for art's sake," or images created in dark caves as a form of "sympathetic hunting magic," designed to ensure success in the chase. Deciphering the meaning behind the art has exercised scholarly minds for generations—producing statistical analyses of the superimposition and placement of the paintings, attempts to decode the development of the art from minute studies of individual friezes and caves, and so on.[12] At the same time, copying methods have come a long way from the tracing paper and acetylene lamps of early workers laboring in dark caves before World War I. Today's researchers use color photography, infrared, and all the technological devices of the digital age. Now we can date individual paintings, as is being done in an increasing number of caves. But we still have a long way to go to decode the intangible meanings behind the art. All we can do is look closely through a cloudy lens, armed with observations of hunter-gatherer art from other societies and oral traditions that record some of the ways in which people used art to communicate with the supernatural.

We think of paintings as finite works of art to be admired and cherished for generations, even millennia. But the Cro-Magnon images were obviously far more than art; they were objects of significance—images with potent ingredients of ocher and blood. We only have to look at living hunter-gatherers to understand this, although obviously the precise meanings of Cro-Magnon paintings elude us. Australian Aborigines painted animals and complex symbols on rock faces as part of their rich ceremonial life, commonly called the Dreamtime. San hunter-gatherers in southern Africa painted for thousands of years. They depicted hunting scenes, moments in camp, dancing men and women, hunters pursuing game, and people cavorting around dying eland. Like the Australian art, this was far more than art for art's sake. Between 1866 and 1874, a German linguist, Wilhelm Bleek, collected oral traditions about rock paintings from San hunters working as convicts on the breakwaters of the Cape Town harbor. Rock-art expert David Lewis-Williams discovered

Bleek's long-forgotten notes and learned from the notes and his own observations how San paintings linked living people with the supernatural. In many San paintings, a human or an animal enters or leaves through a crack, climbs an uneven rock face, or emerges from a shelter wall. Perhaps the rock shelters were entrances to the spirit world, the wall itself being a kind of curtain between the living and supernatural realms. Lewis-Williams also learned that the San both looked at and touched their paintings. Some friezes in rock shelters and caves display wear resulting from repeated hand contact. In other places, rock shelter walls bear imprints of hands that were covered with paint.[13]

What possible relevance can the paintings executed by artists living in a semitropical African homeland have to the Aurignacians and to Grotte de Chauvet? I believe there are some striking similarities *at a general level*, which may have been shared by numerous hunter-gatherer societies. At Chauvet, as in southern Africa, there is a close relationship between painted figures and the contours of the rock. Here, too, visitors to the cave made imprints of their hands on the wall, as if drawing something, perhaps potency, from the rock. Then there are the animal-human figures: the Lion Man from the Swabian Alps, the enigmatic partly human figures from Chauvet, which hint strongly that animals were seen as people and that some members of society were able to transform themselves in their imaginations, and those of their observers, into animals.

Grotte de Chauvet displays a bestiary of dangerous and formidable beasts, with few instances of humbler, easier-to-catch animals. People may have drawn power from the rock face; a solitary owl watches over one of the chambers, in later times a powerful link with the supernatural. No one who visits Chauvet can doubt the complexity and sophistication of Cro-Magnon life. The first modern Europeans added a dramatic new element to human existence: their imaginations and their belief that they lived out their lives surrounded by the living forces of the supernatural world, in the midst of a landscape imbued with profound symbolism. Clearly, these general ideas were very ancient by the time they reached late Ice Age Europe. We know that San beliefs, also of great antiquity, were common to enormous areas of East Africa and southern Africa, and it is not too far of a stretch to assume that notions of animals as living people, of potency acquired from prey, and of a ubiquitous supernatural realm were common to the first modern groups that moved out of Africa over fifty thousand years ago. And in time, these general beliefs may have formed the spiritual underpinning for Cro-Magnon society, even if almost all the details, and, indeed, the animals involved, were totally different.

There has never been a hunter-gatherer society on earth created by *Homo sapiens* that did not possess a complex set of supernatural beliefs or consider itself as living in an intensely symbolic realm. These beliefs and assumptions have nourished small-scale societies constantly on the move, often in times of major climatic change. For instance, the Australian Aboriginal groups living in the central Australian desert adjust constantly to drought cycles. They do so by moving, but with minimal alterations to their tool kit or way of life. What does change, however, is the fabric of spiritual beliefs and ritual observances, which adjusts dramatically to intense drought or other changing circumstances.[14] The people we call Aurignacians spread rapidly across Europe, traversing rapidly changing, often cold and inhospitable environments, as harsh in their way as those of the Australian desert. Perhaps we should not be surprised to find that they enjoyed an elaborate spiritual life, which revolved around the most important reality of their lives: the bestiary that surrounded them from which they acquired a compelling potency and upon which their survival depended.

The Cro-Magnons survived in Europe and Eurasia because they had developed means of locating game and other foods that lay over the horizon, as well as means of communicating this information over long distances. Their projectile technology was highly effective when bolstered by the often long-distance relationships among individuals and fellow kin. These relationships went arm in arm with complex symbolism and intricate relationships with a powerful supernatural world, so brilliantly epitomized by the engraved and painted walls of Grotte de Chauvet. As their new homeland cooled after thirty-seven thousand years ago, to culminate in the Last Glacial Maximum, the Cro-Magnons mastered some of the harshest environments on earth during millennia of unprecedented and often rapid climatic change.

CHAPTER 8

Fat, Flints, and Furs

MORAVIA, LATE WINTER, TWENTY-NINE THOUSAND years ago. The twelve-year-old boy sits listlessly by the hearth, shoulders sagging, seemingly exhausted. He has spent the day in the cold, setting arctic fox traps with his father, with nothing but a few handfuls of lean, dried meat to eat. These are the days of hunger, when the reindeer are thin, when the people have burned off most of the fat on their bodies just keeping warm. The boy's mother takes a practiced look at him, reaches for her precious cache of reindeer fat from the fall hunts, and melts a lump over the flames. She pours the rich lard into a skin container and passes it to her son, who drinks it greedily. He wraps himself in thick furs and drifts off into a deep sleep.

As a cold, gray dawn breaks, the boy awakes, full of energy once again. His father wants him outside to visit the traps, but his mother objects. His boots are too small, and the soles are almost worn through. He fidgets impatiently as she makes him stand on a reindeer-hide laid on the floor. Then she crouches at his feet, cutting the hide round each one with a sharp flint knife. She rebukes him sharply as he shifts his weight from one foot to the other. "We will leave now," says the father angrily as he polishes his spears. But the mother insists on finishing. There's more fitting to be done. She stretches carefully prepared reindeer-leg skins around his legs, brown fur on the inside, cutting and tapering them carefully, piece by piece, deftly stitching at key seams, especially around the ankles. The boots take shape like snug tubes around her son's legs. Satisfied, she tells him to take them off carefully and then hands them to the boy's older sister. The father sighs with impatience and gestures his son outside. They gather their parkas and spears and vanish into the open. Meanwhile, the mother sits beside her daughter, watching carefully. Hesitantly at first, the girl joins the strips of reindeer-leg hide one to another with needle and thread, the forelimb skin at the front and the back-leg skin at the rear. The mother nags and encourages her daughter, checking that the seams are as tight and regular as possible. Then she watches her oversew the original stitching to make the seams waterproof.[1]

Neither mother nor daughter strays far from the hearth on this overcast, cold day, but the hunter and his son are well out in the open. A light wind from the north hints of snow. The snow-covered valley is gray and still under the lowering overcast. The conifers on higher ground stand black and motionless. A solitary bison by the frozen river paws the snow for the lichens below. Oblivious to the biting cold, father and son walk slowly across the level floodplain. Their eyes are never still, always on the alert for stalking predators lurking among the dark boulders at water's edge. The hunters walk for hours without seeing an ermine-white arctic fox flitting over open ground. Hunting such elusive prey requires infinite patience and skill born of long experience.

As they walk, the father gives a commentary on the search, passing on information about foxes learned from hard experience and from his own long-dead father. He recounts how their prey range widely over the nearby plains in summer feeding on berries and small animals. Their summer pelts are drab brown, and they are hard to spot, let alone hunt. Arctic foxes are winter prey. By then, the fur is white, making them virtually invisible against the snow—which is where experience counts. The hunter has set traps close to predator kills that he noted before the snows, for the foxes smell out such places and tunnel under the snow. He reminds his son of a long-abandoned meat cache from an earlier summer prepared by a fox but never consumed. He explains how the foxes urinate in the snow to mark carrion under the surface. Such places are where he sets his traps.[2]

The urine stains are faint, but enough remains to show that a fox has passed nearby. Two short rows of blocks of snow lead to a dark square in the ground. A frantic scrabbling sound reaches the hunters' ears. The son slips ahead and peers into the hole. He calls triumphantly. A fox is impaled on the sharpened antlers at the bottom of the trap. The father kills it with a quick spear thrust. His son lies on the snow and lifts out the carcass. He sets it aside, guts it quickly, and helps set the trap once more. "That's why I want you here," says the father as they reset the flexible sticks that bent, then slipped back into position as the fox fell through. Then they set new bait on the upwind side of the trap between the rows of snow blocks that steer their prey to the tempting morsels, and they move on to the next trap.

At the end of the day, the hunters return home with four trapped foxes. Now the father watches as his son takes a small, sharp-edged flint and makes incisions along the insides of the legs. Next he removes the pelt from them. He then cuts back the skin around the lips before peeling it back over the head, turning it inside out and cutting it free. His sister scrapes the fat off the skins with a stone scraper, then stretches them on wooden frames to dry inside the house . . .

UNBEKNOWNST TO THE boy with his new boots, Europe's climate and, even more so, its geography were changing profoundly.[3] The Cro-Magnon world was getting colder. Snow and frost endured into May or even June. The first freeze came by mid-September, sometimes even earlier. Now winter lasted for well over six months, sometimes as long as nine, especially on the bitterly cold open tundra. Ice cores, pollen sequences, and geological observation throughout the West reveal that snow levels fell and white tides of glacial ice advanced inexorably down northern mountainsides. By twenty-seven thousand years ago, the Scandinavian ice sheet had expanded from virtually nothing to cover all of Norway and Sweden, as well as what is now the Baltic Sea. Glaciers mantled two thirds of the British Isles. Far to the south, the great Alpine glaciers mantled much lower slopes; the foothills of the Pyrenees were an icy wilderness (figure 8.1).

Away from the ice sheets, the landscape was treeless, vegetation sparse. Between forty thousand and twenty-five thousand years ago, insect fossils tell us, mean July temperatures in Britain were close to fifty

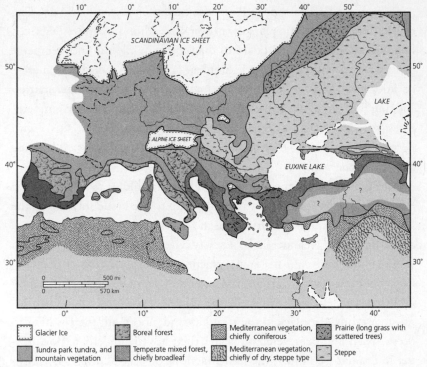

Figure 8.1 *Europe during the Last Glacial Maximum.*

degrees Fahrenheit (10 degrees Celsius)—and that in an area with a generally oceanic climate. A huge expanse of steppe-tundra stretched eastward from Britain and the then-dry North Sea deep into Siberia. During the preceding, warmer millennia, the North European Plain in the West had alternated between coniferous forest with tundra in the north and a mosaic of tundra and cold steppe during colder intervals. Now much of Europe was slowly becoming a polar desert. Pitiless north winds from the ice sheets blew clouds of fine glacial dust over the plains. By twenty-two thousand years ago, sea levels were 330 feet (100 meters) or more lower than today. Much of the North Sea was dry land; Britain was part of the European continent; an extensive continental shelf extended west from what is now France. Everywhere, tree growth was sparse or non existent, except in sheltered locales like deep river valleys north of forty-two degrees north, the latitude of central Italy.

By now, the East European Plain was exceptionally cold and dry. The southern range of arctic plants was over eleven hundred miles (eighteen hundred kilometers) further south than today. While summer temperatures were around 52 degrees Fahrenheit (11 degrees Celsius), a mere 2 or 3 degrees below modern levels, winter readings were another matter. Judging from what we know of the vegetation, mean January temperatures were about −25 degrees Fahrenheit (−3.8 degrees Celsius), with permafrost conditions over a wide area. Winters usually lasted for nine months of the year.

Europe had experienced a similar gradual buildup to extreme cold before, during the millennia that had led up to the first cold snap of the last glaciation, about seventy thousand years ago. The only Europeans at the time were Neanderthals. They survived by moving into sheltered valleys and slightly warmer environments in Italy and south of the Pyrenees, but even during warmer millennia they never settled permanently in the river valleys on the open steppe-tundra of the North and East. Almost certainly, they lacked the tool kits and, above all, the clothing to do so. Their numbers were minuscule, their hunting mostly opportunistic. In dramatic contrast, the Cro-Magnons of central and eastern Europe flourished when confronted by similar increasing cold thousands of years later. Their numbers increased significantly in what was slowly becoming one of the coldest environments on earth.

Why did the Cro-Magnons succeed under circumstances where the Neanderthals retreated?[4] We can point to their superior cognitive skills, to their fluent speech and ability to cooperate and plan ahead, to their rich spiritual life, but these were already in place thousands of years earlier with the arrival of the Aurignacians. What really came into play with the onset of the cold were the qualities of restless innovation that had marked *Homo sapiens* from the beginning. We tend to forget just how recently the ancestors of the Cro-Magnons had left the open spaces of Africa and southwest Asia, environments where they had hunted over large territories with both game and plant foods dispersed over long distances. The first Cro-Magnons to enter Europe and Eurasia still owed much to their African ancestry, to hunting skills and lighter weaponry adapted in open, somewhat arid landscapes. They were fleet of foot, clever stalkers, and adept at using the landscape to ambush game, skills that would serve the Aurignacians well during the changeable, somewhat warmer conditions before twenty-nine thousand years ago. Wild plants and fruit were unimportant in landscapes where growing seasons were extremely short. Judging from a study of modern Alaskan caribou hunters, most people probably consumed no more than a cupful or so of plant food a year.

Despite some short warmer intervals, winters over much of the continent were long and savagely cold, especially in central and eastern Europe. It was here, in shallow river valleys and compact settlements dug partially into the soil, that Cro-Magnon groups refined the art of hunting in cold deserts, in vast landscapes, and on the margins of coniferous forests to adapt to a much colder, often-treeless world. For thousands of years, their ancestors had survived thanks to their mobility and by making use of small, highly portable tool kits and relying on multipurpose tools for a wide variety of tasks. People living in the Danube Basin, in the valleys at the edge of the northern tundra and on the seemingly endless expanses of the East European Plain, adapted effortlessly to the ever-colder conditions. The cultural changes were imperceptible from one generation to the next, but cumulatively they resulted in the world's first truly arctic-adapted cultures. The essentials of survival remained unchanged: meat, of course, but also, and just as important, fat and furs, needles and thread.[5]

CAUTIOUS, INNOVATIVE, AND absolutely versatile, the diverse Cro-Magnon societies of the Last Glacial Maximum were an exemplar of all later arctic hunter-gatherer societies. The number of cultural options available to people living in extremely cold environments are necessarily limited by the realities of long winters with very low temperatures, strong winds, snow, and ice.

Obviously, everyone was well aware of the hazards that confronted them: overheating, frostbite, and fat deprivation caused by eating too much lean meat. Fat provided vital calories and was a central part of their diet. Cro-Magnons lived in environments where energy sources were in short supply during the winter months. Fat is a source of energy and heat when food is scarce. Birds and mammals, for example, use more than 85 percent of their dietary calories to maintain a constant body temperature. A genetic ability to fatten rapidly is an adaptive response to scarce times of the year. Many animals use fat as a way of transferring energy from the season of abundance to that of scarcity. Most mammals living in late Ice Age Europe and Eurasia needed large fat reserves to survive the tough conditions of the winter. Winter saw slowly declining body weights that could sometimes assume lethal proportions. Fortunately for most animals, the arrival of spring arrested the decline. Were Cro-Magnons any different?

In human terms, any diet of lean muscle has few calories, about one thousand per 2.2 pounds (1 kilogram) of dry weight.[6] Hunters working hard outside in cold environments could easily metabolize about four thousand to five thousand calories a day, equivalent to about 9 to 11 pounds (4 to 5 kilograms) of lean dried meat. That would be just enough to maintain their weight, let alone put on more poundage. Fat was only available from their prey for a relatively short time each year, in late summer and early fall, when reindeer and other animals had accumulated massive fat deposits. This is why late-summer and fall hunts were so important in Cro-Magnon life and probably why many of the animals depicted on cave walls appear with swollen bellies, with the massive fat deposits that carried them through winter. At the time of the fall hunts, everyone, animal or human, was searching for fat. Thus, argues paleontologist R. Dale Guthrie, it would be hazardous to keep much of it around one's camp for fear of attack. He believes that people put on weight by eating large quantities of fat as they acquired it so that they, like the animals, could survive the lean months ahead.

The Cro-Magnons confronted the same limitations and responded in full, just as historic Eskimo and Inuit societies did into the last century, and do even today. This is why we can learn many lessons from these societies about the hunting methods and behavior of late Ice Age people, who lived in environments as harsh as, if not harsher than, those of the modern world. After all, the number of ways in which one can kill an arctic fox without firearms is necessarily limited. As we saw with the boy and his father, such techniques depended on close observation of the prey and its habits and on a realization that trapping was the only effective way to obtain their furs.

Even in warmer years, everyone lived in environments where temperatures could change rapidly, where timber was often in short supply, and where frostbite and hypothermia were daily hazards, as was, during the short summers, the opposite: overheating. To survive and hunt effectively in these shifting temperatures, they needed close-fitting, layered clothing. To create it required not only numerous furs but also a means of sewing them: the needle. But they could not manufacture needles without "unlocking" the only raw materials in abundant supply in often deep-frozen environments: antler, bone, and ivory. By thirty thousand years ago, Cro-Magnon toolmaking philosophy for both antler and stone had become remarkably like that behind the modern-day Swiss Army Knife.

MUCH OF CENTRAL and eastern Europe was essentially treeless during colder times. Temperate and tropical hunter-gatherers relied heavily on wood of all kinds for spears, clubs, and all manner of other artifacts, as did much earlier Europeans. Apparently, there was still timber for spear shafts, but straight-grained wood must have been in high demand. The fire-hardened thrusting spear of earlier times, with its limited range, was now an anachronism among Cro-Magnons, who relied on composite, lightweight weapons that could be hurled accurately over distances that approached fifty feet (fifteen meters). The effectiveness of the new technology depended on lethally sharp stone and bone projectile points and, as time went on, spears armed with small, razor-sharp stone barbs.

Just like the Neanderthals, the Cro-Magnons counted on their knowledge of the habits of the bestiary and on their brilliant stalking abilities. Even with spears without throwers that had a range of fifty feet (fifteen meters), the hunter had to approach his quarry close enough to at least partially immobilize it or wound it seriously, so he could track it as it weakened and died. Unlike bows and arrows, spears cast by hand lack much velocity, even in expert hands, unless propelled by hooked artifacts known as spear throwers.

The earliest known spear throwers come from French rock shelter levels dating to about eighteen thousand years ago. These are bone tools, but that does not necessarily mean that wooden throwers were not in use far earlier. Spear throwers would have been very effective on the tundra and in situations where people ambushed a small herd of animals like horses or reindeer. Like needles, spear throwers were vitally important artifacts. A hunter can throw a light spear fast and far, but there are limits to the speed at which a human arm can move. Spear throwers are basically an extension of the arm, and they lengthen its throwing reach by exerting additional force on the spear. The weapon has increased range and velocity, though only if it's lighter than the conventional arm-propelled spear. It now flies with a flatter trajectory and usually has greater penetrating and killing power. Not only that, but in flying faster, it spends less time in the air, which translates into greater accuracy and less time for the prey to move. In its simplest form, the spear thrower was little more than a hooked stick, the spear fitting into the hook and lying along the thrower.

But there was far more to the Cro-Magnon spear thrower and its dart than just a hook. Experiments have shown that, like bows and arrows, the weapon had to be tailor-made for its owner. The feel and balance of the spear when hooked on the thrower varied from person to person and required meticulous adjustment based on the cumulative experience of generations. Everyone fine-tuned their throwers by weighing or shaving them, by shortening and balancing the spear. Weight was an important variable. Modern experiments show that the longer the spear thrower, the shorter the spear, but there must have been many variations depending on the game being pursued and hunting strategies in use. When using a modern replica of a North American spear thrower, I noticed that a good thrower flexed slightly when in use, a built-in spring that released additional power at the end of the cast. When I used a weighted thrower, the effect was enhanced, which may be why later Cro-Magnon throwers bear elaborately carved images of animals that have the effect of serving as weights (see color plate 10). (The tradition of elaborate decoration lingers today in beautifully crafted, and expensive, shotguns.) Unlike projectile points, spear throwers were probably valued possessions that remained in use for a long time, which may also explain why many of them were elaborately decorated. For the most part, however, Cro-Magnon throwers were probably simple devices, easily made, light to carry, and, in skilled hands, stunningly efficient hunting weapons. Some of the smaller spear throwers found in later sites may be ones used by children, who surely learned to use these weapons at an early age.

In a treeless environment, finding raw materials to create refined artifacts like spear throwers required ingenuity. In the mostly treeless environments of the north, people turned to antler and bone, not only for tools and weapons but also for building materials and fuel. No one could have hunted, let alone settled, on the open steppe-tundra without ready access to the most plentiful of all raw materials: animal skeletons. To have such access, however, they needed highly effective stone tools to turn antler and bone into fine, razor-sharp weapons and other small artifacts. This meant unlocking the potential of mammoth ivory, reindeer antlers, and long bones. Without lengthy splinters of such materials, no Cro-Magnon could create versatile, lightweight weapons or such delicate artifacts as bone awls and that most revolutionary of inventions, the eyed needle. The punch-struck blade provided the blanks for such artifacts. Only the ingenuity of the maker limited the design of what were multipurpose tools made from dozens of relatively standard-sized blades. This simple but versatile technology resembled the aforementioned modern-day Swiss Army Knife: a chassis that supports a variety of tools (figure 8.2).[7]

The most important product of the Cro-Magnon Swiss Army Knife was the burin, the chisel of the late Ice Age, which cut through the tough outer surfaces of antler, bone, and mammoth ivory (figure 8.3).[8] Burins are not much to look at, remarkable only for the short longitudinal flakes removed from their ends to produce the chisel edge. They take many forms. Some have simple, one-sided edges. Others are double-ended, some even beaked. Classifying burins tends to drive archaeologists crazy, for they defy precise ordering into neat categories. They were, however, among the most important artifacts ever fashioned by Cro-Magnon stoneworkers, who followed some basic forms but varied them according to the blades and tasks at hand. Not only were burins

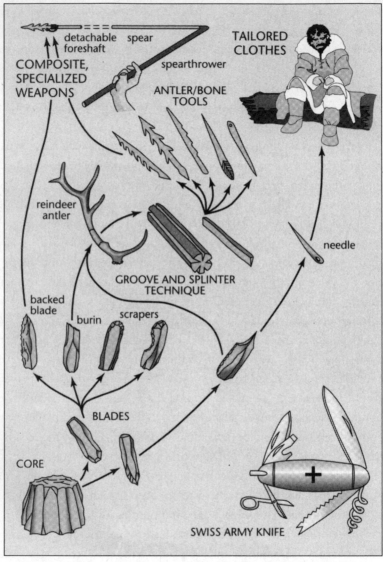

Figure 8.2 *Swiss Army Knife technology and the Cro-Magnons' light, composite weaponry. Much of Cro-Magnon technology depended on relatively standardized blades, struck-off cores of fine-grained rock. The core was the "chassis" like that of the Swiss Army Knife. A modern user simply unfolds the artifact attached to the knife body. Many Cro-Magnons probably carried stone cores around with them and struck off artifact blanks when the need for a specialized tool arose.*

Blade technology, or the Cro-Magnon Swiss Army Knife

Like so much else about Cro-Magnon society, using blades as blanks for multipurpose tools goes back thousands of years into the African past. As we have seen, blade technology first appeared in Africa shortly after seventy thousand years ago and was widely used in the Near East and Europe by forty-three thousand years ago. Almost invariably, these early blade makers used stone hammers to strike off large, relatively standardized blanks. Eventually, they refined the technique. Before temperatures cooled around thirty thousand years ago, Cro-Magnon stoneworkers were using antler and bone punches to strike off remarkably standard-sized, thin blades from carefully shaped conical cores. To do so, they required high-quality toolmaking stone, which was one reason why hunting groups in central and eastern Europe acquired such material from sources far over the horizon.

We know a great deal about the techniques of blade making, thanks to modern-day experiments by expert knappers, or lithic technologists, as they prefer to be called. I tried for days to master blades at a lithic technology field camp, but never managed to fabricate anything except a few crude flakes. More adept practitioners like the late Don Crabtree of Idaho and Jeff Flenniken of Washington State could, or can, turn out conical cores and dozens of astonishingly standard-sized blades in minutes. By watching them, I learned that the raw blade is a highly effective multipurpose tool on its own, with a cutting edge so sharp that Crabtree fabricated some blades from obsidian that were then used by the surgeon who performed eye surgery on him. It is a matter of gaining the basic skill, and the habit, of acquiring a feel for the essential properties of a lump of toolmaking stone before it is even shaped into a core. Cro-Magnon stone workers could detach the blade with a simple antler punch and hammerstone or with a wooden chest pump with the antler tip placed atop the core. Either way, they could produce a single blade, or half a dozen, in a few seconds.

Each conical core produced dozens of blades, all of them potential blanks for all manner of specialized tools. Unlike Neanderthal artifacts, or even many of those produced by Aurignacians, these awls, burins, knives, projectile points, and scrapers served highly specific purposes. They were also versatile enough to serve as multipurpose tools, perhaps as projectile points or skinning knives, within the space of a few minutes. Fine awls drilled holes in needles; notched and serrated blades smoothed antler, bone, and wood. Steeply backed, needle-sharp blades formed the tips of hunting weapons and also served as delicate knives.

The Les Eyzies Museum is one of the best places to explore the bewildering range of Cro-Magnon stone and bone tool technology as it developed over thirty thousand years. The displays are a testimony to the economical way in which Cro-Magnons used stone and bone. It was here that I first thought of the Swiss Army Knife or Leatherman of today in relation to Cro-Magnon tools. Both are basically a chassis with strong hinges, attached to which are all manner of specialized tools: knives, spikes, nail files, saws, even scissors. I realized that the Cro-Magnons had the same idea with the carefully shaped blade core. This was the chassis of their Swiss Army Knife, the basis for numerous small tools, almost all of them made by slightly trimming or backing a blank from the core. A Cro-Magnon who carried around a flint core and an antler punch could readily produce any tool he or she wanted with a quick punch stroke, then some deft retouching. A spear stave needed smoothing, and a notched blade served as a spokeshave; a projectile point broke, and a replacement was available in a few minutes. The punch-struck blade was a wonderfully versatile blank for all kinds of artifacts. Only the ingenuity of the maker limited the design of what were, effectively, multipurpose tools.

versatile; they were also the artifacts that grooved and broke open fresh reindeer and red deer antler—the doorway to an entirely new technology. The same tools also worked on mammoth ivory, especially if a tusk fragment was soaked in water. Artists used burins to engrave bone and cave walls, too.

The burin was fundamentally a grooving tool, which cut through the deep ridges in fresh antlers or, and this was more work, produced slivers of limb bone. How effective was it? The French prehistorian André Rigaud grooved reindeer antler with a dihedral (double-faceted) burin. He cut longitudinal grooves in a fresh antler beam, curving them, so that they crossed one another at one end. The sharp flint penetrated the hard external surfaces, which require a sharp, hard edge to cut through longitudinally.[9] A large reindeer antler beam would produce several splinters, acquired by cutting through the hard exterior into the spongy interior. The toolmaker then drilled a hole with an awl through the soft interior matrix, pushed through a stout thong, and levered out the splinters one by one. The blanks were like stone blades, easily fashioned into points or other weapons. Bone was tougher to work. Judging from historic practice, the joint ends of a long bone were removed, and the shaft was soaked for a day in water, which softened it. When the bone was softened, the toolmaker wrapped his or her fingers in a skin around the burin, drawing the tool carefully toward him or her. The bone got hot and gave off a pungent smell as it was grooved. A scratching noise filled the air. When the worker thought the groove was deep enough, he or she tapped it, and the splinter fell out.

Once the burin key is turned, you can think of a second form of Cro-Magnon Swiss Army Knife, this time an antler rack, the chassis for all kinds of plaques, slivers, and beams that then became spearheads and harpoons, needles and other small objects. Without the two Swiss Army Knives of Cro-Magnon technology, manufacturing truly lightweight weaponry on the steppe would have been near-impossible.

The most revolutionary artifact of all produced by a burin stroke was the eyed needle (see figure 8.3b). It's no exaggeration to say that this seemingly unimportant device changed history.

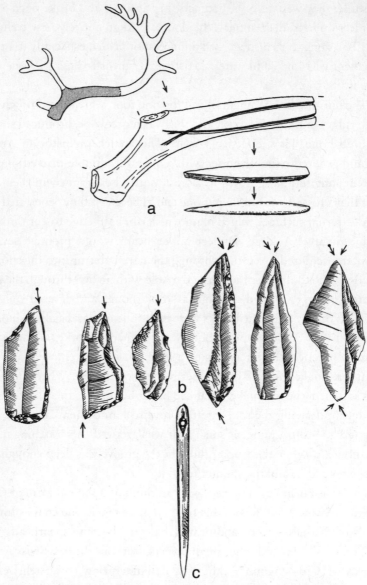

Figure 8.3 *(a) A reindeer antler grooved with a burin to produce splin-*
ters as blanks for antler points. (b) Various forms of late Ice Age burins.
The arrows show the chisel edges. (c) A Magdalenian eyed needle. (a)
and (b) after Jean-Luc Piel-Desruisseaux, 2007. Outils Préhistoriques
(Paris: Dunod). Reproduced with permission. (Copyright) Brian Fagan.

THE HUMBLE NEEDLE ranks alongside the taming of fire as one of early humanity's most significant innovations. For tens of thousands of years, Neanderthals and their predecessors had draped skins around their bodies like capes as protection against the cold. As, presumably, the Neanderthals well knew, you can take a length of fiber or thong and join two hides with a seam by pulling the "thread" through holes made with a flint awl. The resulting garment is rough at best, perhaps a tubular pair of pants. Needles allowed women to make formfitting, layered garments.

Modern-day adventure outfitters sell layered protection from the elements as if it were a startling innovation. In fact, thanks to the needle, the Cro-Magnons were making remarkably similar layered clothing, albeit from different materials, over twenty-five thousand years ago. For the first time, women could fashion garments tailored to size for infants, children, and growing youths, as well as for adults and old people. They could also sew clothes of all kinds: underwear, light shirts and long pants for summer days, and the thick fur parkas and lower garments worn in winter. The sewers used hides and furs from different animals to make composite garments that maximized the protective qualities of different species. For instance, modern Eskimo hunters know that two reindeer hides sewn together with the thick fur inside make an excellent sleeping bag, heavy but highly effective in temporary winter camps.[10] And we can be sure that late Ice Age people used them as well. Not that the Cro-Magnons confined themselves to furs and hides. Presumably, like historic Inuit and Aleut tailors, they made use of such exotic materials as bird skins and seal stomachs, each of which had uniquely useful qualities for such prosaic items as sleeping bags.

As historic Eskimo, and undoubtedly Cro-Magnons, well knew, even the best skin clothing was insufficient to protect stationary hunters on subzero days. The hunters would have been careful to keep moving, even running to keep warm. When lying in wait for game, they would have stamped their feet, slapped their hands together, and otherwise taken precautions to avoid frostbite and dangerous chilling. By the same token, during warmer months, it would have been easy to become overheated while hauling meat or engaging in other hard work outside. For much of the year, lighter parkas would have been most effective, with heavier layers underneath. In this way, they would have avoided over-heating that would have trapped sweat and, eventually, caused chilling. Adding or shedding layers of clothing, combined with commonsense precautions against hypothermia, would have allowed them to maintain a comfortable level of warmth whatever the temperature. A lack of tai-lored, layered garments may have been one reason why Neanderthals rarely settled for any length of time on the North European Plain.

There are two potential resources that allow us to speculate intelli-gently about Cro-Magnon garments. Unfortunately, burials, a potential primary source, reveal few details of clothing, except for the spectacular interments of the Sungir people, described in chapter 9. As with hunting methods, we can also fall back on accounts of traditional Eskimo and Inuit clothing, for we can be sure that the demands of subzero environ-ments meant that basic clothing styles in such regions changed but little between late Ice Age times and the nineteenth century. In historic times and in the remote past, everything depended on the unique qualities of different hides and furs, which never changed over the millennia.

Judging from historic arctic societies, the most important garment was the hooded parka, usually made from reindeer skin.[11] Parkas would always have been worn with the fur on the inside and the skin outward. The deep hood would have had a fur trim, made from the pelt of an an-imal, which readily sheds snow. A belt would have prevented heat loss at the midriff. Such parkas would have been exceptionally warm when combined with long pants with the fur also turned inward. The Cro-Magnons may have also used lighter pants, for reindeer ones would have been too warm in temperatures above zero and been somewhat restric-tive of movement. Perhaps their protection extended to facial adorn-ment. The anthropologist Richard Nelson observed that many Eskimo in Wainwright, Alaska, used to grow mustaches each fall, which they kept until spring. The mustache hairs collected frost rather than ex-posed flesh, and the fresh ice that accumulated there could be used for water in an emergency.

Hunting in subzero temperatures required Cro-Magnon men to work outside for hours on end. Even on longer, warmer days, temperatures could remain dangerously low. They must have been well aware that you lose heat from your extremities, which made head protection, gloves, and footwear of great importance, even during the brief summers. Here traditional Eskimo boots give us a clue as to what the Cro-Magnons might have worn.[12] The sole was the most important part of the boot, and it may have been made by cutting sections from a skin, from which the fur was removed with scrapers, or by leaving it outside to rot during the summer. A sole could be slippery, so the maker may have sewn some skin strips across the toe and heel. The uppers may have been made from the skin of reindeer legs, from a beast killed in fall, when it was in prime condition. Skinning the legs would have taken care, as the maker slit the forelegs down the front and the hind legs down the back. Once dry, the hide would have been scraped until it was soft and pliable on the inside. The four legs would have been sewn together with braided sinews to form the uppers, which would have reached to just below the knees. A sinew drawstring at the top would have kept out snow.

Reindeer-skin insoles and long socks with the fur inside would have kept the wearer warm, with the added advantage that they would never freeze into the boot. This kind of footwear is very effective provided that the wearer takes care to dry everything. Drying boots and socks must have hung from the roofs of every Cro-Magnon dwelling.

The reindeer, with its short fur, would have provided the raw materials for gloves. Simple mittens made of reindeer, wolf, or even bear skin would have sufficed in many conditions. Hunters on the move would have hung their gloves or mittens around their necks to avoid losing them.

THE FOX HUNTER and his son, whom we met earlier in the chapter, were completely at home amidst the snow and ice, not only because of their layered clothing and efficient technology but also because they made use of a battery of routine skills acquired since birth from their elders and their own hard experience. The expertise they needed was, in many respects, identical to that used by their distant African and Near Eastern ancestors: the ability to track prey from the most subtle of indicators far beyond foot imprints, remarkable skill at stalking at close range, and, of course, the capacity to inflict serious wounds or to kill with relatively simple weaponry. Infinite patience and persistence were also qualities common to tropical and cold-climate hunters alike. Everywhere, mental attitudes were important, but they were particularly central to survival in environments where strong winds and the bitter cold of subzero temperatures for days on end sapped human energy. Successful hunting and survival depended on deeply ingrained attitudes of self-assurance and competence, on mental attitudes that were part of the Cro-Magnon personality.

We cannot, of course, decipher these day-to-day skills from archaeological sites or cave art. Fortunately, anyone living in arctic and periglacial (areas near ice sheets) environments requires the same constellation of human qualities. Richard Nelson's *Hunters of the Northern Ice*, published in 1969, is a classic account of the challenges of living in extreme environments.[13] I make no apologies for extrapolating his analysis of Eskimo behavior back into the Cro-Magnon culture. There are few other ways of surviving in a subzero world.

Like Nelson's Eskimo, every Cro-Magnon, whether man or woman, must have possessed a near-encyclopedic knowledge of the environment at all seasons of the year that extended far beyond the normal routines of travel and subsistence. It would have covered unusual occurrences, like a sudden snowstorm or drop in temperature, being trapped by a flood during the spring thaw, or a hunting injury. We would call many of these situations "emergencies," which often leave us at a loss. Judging from Nelson's research, Cro-Magnons would have responded to them instinctually, to the point that they would not have been emergencies at all. For example, hunters would have moved immediately to make a camp in as secure a place as possible if a snowstorm had approached without warning, or moved to higher ground if a sudden thaw had brought flooding to a riverbank where they were hunting. They knew the telltale signs of impending danger and acted upon them.

Like all modern hunter-gatherers, whatever their homeland, the Cro-Magnon hunter would have learned as much as possible from others, especially experienced men, who were respected for their profound knowledge of the chase. Nelson recounts how the Eskimo would watch each other hunt, then discuss their experiences later. These personal observations, acquired through tough experience and by listening, would always have been shared with others among the Cro-Magnons. In this way, the cumulative experience of the group passed from one generation to the next. This was essential in a world where life expectancy was short and where a sudden hunting accident could decimate a band in a few minutes.

Nelson observed that every Eskimo was self-assured about his or her knowledge. The same must have been true of the Cro-Magnons. Their survival, and success in the food quest, depended on such self-assurance and on deeply ingrained qualities of both mental and physical perseverance. Everyone would have been in excellent physical condition, able to perform difficult tasks hour after hour without faltering. Physical stamina was all-important, but mental toughness was just as vital: the ability to withstand physical discomfort such as cold, wetness, or intense exertion without being constantly concerned about one's comfort, as we are. Cro-Magnons traveled light with a multipurpose tool kit, without all the impedimenta of food and clothing that so many of us regard as essential today. This not only lightened the load but also allowed for more space for carrying prey. The hunter would have taken the minimum of equipment other than that needed in an emergency, especially during the winter months, when cold was always a factor and survival depended on both experience and the ability to respond rapidly to unexpected circumstances.

The same qualities of persistence must have extended to every journey or hunt, for such trips, once begun, would rarely have been cut short. In Nelson's experience, the secret was extremely thorough preparation, careful assessment of potential dangers and observation of weather conditions, and the expectation that every task, however hard, would be completed, however long it took. The completion of a task did not necessarily mean squandering energy, especially on the move. Experience taught that the easiest way through rough terrain was not simply to cross it but to reconnoiter carefully, then take the smoothest way ahead, even if it involved a longer distance. Almost invariably, the

Cro-Magnons would have used every intelligent shortcut to save unnecessary effort.

Perseverance would have gone hand in hand with foresight, especially during hunts or in other situations where there was immediate potential danger. Like modern-day arctic hunters, the Cro-Magnons would never have taken unnecessary risks like walking on thin river ice. They would have assessed the risks behind every activity and would never have taken the kinds of chances that people do today for the excitement of it.

Alertness goes with foresight, an acute awareness of one's surroundings, even when preoccupied with a single, absorbing task. You could not afford to be distracted by a single activity in a world where predators abounded and the hunters were also the hunted. A hunter setting a trap must have looked up from his work constantly, watchful for a lurking bear or lion. A woman gathering nuts in the trees was always alert for a hungry beast stalking her. Hunting, harvesting plants, fishing, and every other activity, even gathering firewood, was part of a multitasking, which could not only save a life but also sometimes bring extra food to the table.

The food quest for historic northern societies was no mechanistic process, nor would it have been for a Cro-Magnon band. Working in the field required both ingenuity and improvisation. Repairs to weapons were a constant reality: relashing an antler spear point, sharpening a flint knife, cutting a new wooden shaft from a nearby tree at a moment's notice; these were routine tasks. Beyond regular fixes, everyone relied on an ability to improvise that was inculcated into them from birth. The classic, and oft-quoted, example is that of arctic hunters who have been known to use frozen meat to replace sled runners or even serve as a complete emergency sledge. All the resources of the natural environment were at hand. Solutions to even quite complex problems came from a flexible, can-do mind-set where there were no limits or conventions.

No one worked alone, for cooperation when hunting, plant gathering, or traveling was fundamental to Cro-Magnon life, just as it has been in every arctic society. Everyone labored together and helped one another, even with the most trivial tasks, not for the sake of thanks but because the entire fabric of life was based on reciprocity: mutual obligation. Everyone shared the spoils of a hunt and cooperated in an emergency. This ability to cooperate was essential in a high-risk environment, where reacting well to sudden temperature shifts could mean the difference between life and death.

All of these qualities would have served the Cro-Magnons well in a world where climate change was often rapid, winters were severe, and periods of warmth and abundance were usually short. Of course there were exceptions to the norm: quarrelsome individuals, volatile family situations, interpersonal violence, and other far-from-ideal circumstances— but in general, the personal qualities shared by the Cro-Magnons, and their encyclopedic knowledge of their world, were their most powerful weapons for survival. Effective technology, an acute self-awareness, and an intimate relationship with the environment made the Cro-Magnon personality practically invincible.

CHAPTER 9

The Gravettians

MORAVIA, EARLY SPRING. THE PTARMIGAN croak loudly as they feed on lichens among the rocks at water's edge. Wide swathes of pale blue sky on a rare sunny day twenty-seven thousand years ago compete with the roiling white clouds borne on the strong north wind. Great flocks of the white-feathered birds zoom close to the fast-flowing, ice-strewn river, but they really prefer to walk, practically invisible in the snow, grubbing for food among the dark boulders by the water. Crowds of them walk upstream, picking their way across the still-firm snow, feathers ruffled by the wind. Some distance away from the water, two women stand motionless, slightly upwind, watching the birds. The day before, they built a simple fence of reeds parallel to the river with tempting gaps in the middle, laced with bait. Ever-inquisitive, the ptarmigan investigate the barrier, pecking amid the sand and snow at its base. A bird pokes its head into the defile, then starts through, neck forward, heading for the bait. There's a quiet snap. A noose slips tightly around the ptarmigan's neck. As it flutters helplessly, one of the women kills it with an antler club and sets the snare anew. The rest of the birds start up in fright, but soon return as the women resume their watch.[1]

Their settlement is close by, on slightly higher land. The domelike houses squat close to the ground, virtually invisible in the snow except for the thin smoke trails whisked and swirled by the ceaseless wind. Piles of abandoned mammoth and reindeer bones swept clear of snow by the gusting currents lie downwind of the dwellings. Two men sit on a boulder by one of the houses. Thick piles of stone flakes accumulate at their feet as they shape cores and punch off thin blades. One of the stoneworkers selects a long blade and blunts one side with an antler fragment. He holds it up to the sun and inspects the tip. Soon he will mount it on the fresh spear shaft lying against another boulder nearby . . .

EVEN AT THE height of the Last Glacial Maximum, there was always a corridor between the Scandinavian ice sheets to the north and those of the Alps to the south. Important lowland defiles connect the North European Plain in Poland with the Danube Valley in Austria to the south. This area of varying topography, of open plains, sheltered valleys, and mountain foothills, was a place where hunting bands with different traditions met and mingled, where people maintained connections with other groups living long distances away on the edges of the steppe-tundra and in shallow river valleys that dissected the vast plains.[2]

The North European Plain was the northern frontier of the Cro-Magnon world. Rolling tundra swept eastward far over the horizon, as far as the Ural Mountains in what is now Russia. This was a brutal, cold world of gray-brown dust and wind, of unrelenting aridity. And yet, harsh as it may have been, the plains supported a surprisingly rich bestiary, including mammoths, musk oxen, horses, bison, and reindeer— and the tough hunting bands that preyed on them. The tundra and its environs were the crucible where truly arctic-adapted societies first developed in central and eastern Europe after forty thousand years ago. We archaeologists call these people Gravettians, after La Gravette rock shelter, in France's Périgord, where their characteristic stone artifacts were first identified, a generic term that masks all kinds of local variations (see the box "Aurignacian, Gravettian, Solutrean, Magdalenian . . ." in chapter 6).[3]

Almost certainly, Gravettian societies first developed out of Aurignacian roots somewhere in central Europe or western Eurasia, where the climate deteriorated more rapidly than in the more temperate West. It was here, on the edges of the tundra, that the most important technological innovations came into play. Needles, layered clothing, and other cold-climate artifacts and practices seem to have percolated westward to Cro-Magnon bands in southwestern France and elsewhere as the cold intensified.

The Gravettians' eastern homeland is loess country, rolling plains covered with dense layers of windblown glacial dust carried by frigid north winds. Loess (the word comes from the Swiss German word *lösch*, meaning "loose") blanketed thousands of square miles of central Europe, a landscape where there were no inviting caves or rock shelters. In many places, the only refuge from wind and dust lay in shallow river valleys or in mountain foothills. So the people dug into the soil for winter shelter. They carved out oval hollows, then roofed them with the only available raw materials: mammoth bones, hides, and sod. Covering the floors with furs most likely helped make the houses snug in subzero temperatures. Their more-permanent dwellings were dome shaped, little more than low mounds that offered no resistance to the wind yet were relatively well insulated against the cold. The winters were so cold that every hunting band probably remained at the same location for months on end.

THE GRAVETTIANS HAD an obsession with mammoths, or so it seems from their major sites, where elephant bones abound, sometimes in great piles close to houses, often carefully stored in pits dug into the permafrost, and sometimes partially burned in abandoned hearths (figure 9.1). The mammoth was a vital resource, not necessarily as food but for fuel and shelter in cold, often sparsely forested landscapes. For this reason, many Gravettian bands camped habitually at strategic locations overlooking river valleys used by migrating elephants and other animals. They also sited camps near blind valleys, gorges, and other natural traps where game could be surrounded or mired. What we don't know is whether they habitually killed significant numbers of mammoth or merely scavenged their carcasses. If they did hunt them, it must have been with ambushes, just as they may have done at Milovice, in a side valley of Moravia's Pavlov Hills. Here, archaeologist Martin Oliva excavated the massed bones of about one hundred young mammoth associated with large and small Gravettian projectile points, killed about twenty-two

Figure 9.1 *Mammoth drawing from the Grand Plafond, Rouffignac, France. Photo: Marie-Odile and Jean Plassard.*

thousand years ago.[4] Whether this was an actual hunt, or a place where young beasts drowned in floodwaters or, repeatedly over many years, became bogged down and were then killed or butchered, we don't know. The only analogous findings come from Siberian mammoth "cemeteries."

I suspect that the swampy valleys inhabited by the Gravettians in Moravia and elsewhere also had their share of now-vanished mammoth bone beds, which rotted away as temperatures rose after the Ice Age, leaving no trace behind, except in places where waterlogged layers, weathered loess, or peat have preserved them. The hunters had no reason to pursue such dangerous beasts when they were most useful for houses and fuel. All the Gravettians had to do was mine natural bone accumulations from rotting carcasses, which were the equivalent to outcrops of toolmaking stone. Every summer, people would harvest jaws, long bones, and other useful parts from the dead mammoth; carry them home; then store the body parts in pits dug into the permafrost or in water before using some of them for their dwellings. Storing them in these ways kept the bone fresh and more easily workable. They also treasured the great tusks, which, when soaked, could be split open or carved.

SNUG, LOW-LYING, OVAL-SHAPED houses dug partly into the ground—the more durable Gravettian dwellings follow the same general design

Mammoths and mammoth cemeteries

Mammuthus primigenius, the woolly mammoth, is an iconic Ice Age
beast. With their long hair and huge tusks, these arctic elephants
were an imposing sight, especially in a winter blizzard, dark, motionless in the driving wind and snow.[5] Mammoths evolved in Siberia, at
least as early as seven hundred thousand years ago, and thrived on the
rich vegetation of the treeless steppe-tundra. They fed there during
spring and summer, but may have migrated southward into river valleys and other locales where fodder was more plentiful during fall
and winter. In many respects, mammoth behavior probably resembled that of modern-day elephants. We know a great deal about their
appearance and diet from spectacular deep-frozen mammoth finds in
Siberia's permafrost.

In 1901, reports reached Saint Petersburg of a strange "devil creature" half buried in arctic mud at Berezovka, in remote northern
Siberia. The creature was in fact a deep-frozen woolly mammoth
that was so well preserved that the hair and hide, as well as much of
the flesh, had survived. The czar himself financed an expedition to
recover it. The leaders, zoologist Otto Herz and geologist Eugen
Pfizenmayer, found the almost-intact beast in a kneeling position, as
if it had fallen into a gully and failed to extract itself. As they excavated the stinking carcass, which smelled "like a badly kept stable
heavily blended with . . . offal," they fed some of the meat to their
dogs, who ate it with avidity.[6] Herz and Pfizenmayer carried samples
of the carcass on a three-thousand-mile (forty-eight-hundred-
kilometer) sled journey back to Itkutsk and then by rail to Saint
Petersburg during a bitterly cold fall and winter. The well-preserved
Berezovka mammoth caused an international sensation.

Not that the Berezovka find was that unusual except for its excellent
preservation. A Siberian mammoth expert, the late Nikolai Vereshchagin, estimated that thousands of skeletons await discovery in Siberia.
He calculated that almost fifty thousand mammoth tusks were found

along the 600 miles (965 kilometers) of the Arctic coast between the
Yana and Kolyma rivers alone between 1660 and 1915, almost all of
which vanished into the insatiable maw of the ivory trade. Some pale-
ontologists believe there may be more than five million mammoth carcasses buried in the Siberian deep freeze, many of them packed
together in the same locations, leading people to talk of "mammoth
cemeteries."

Mammoth graveyards or cemeteries are a myth that refuses to die.
Bone accumulations do exist, but they are matters of chance rather
than deliberate choice by tottering beasts. There are dried-up river
courses and water holes in both Africa and Asia where elephants
have died repeatedly over many generations, often because of
drought or flood, but they certainly weren't cemeteries.

Vereshchagin's excavations at Berelekh, in northeastern Siberia,
document one such place.[7] A 6.5-foot (2-meter) layer of frozen peat
and icy silt extends at least 590 feet (180 meters) along a riverbank.
Soil movements and water erosion have exposed an estimated fifty
thousand bones from 200 mammoths over the past century alone,
this from a bone bed that extends an unknown distance into a hill
slope along the bank. Vereshchagin and his colleagues used a pump
and fire hoses to peel away the frozen muck, exposing bones as well
as brown and golden mammoth hair. The excavations uncovered the
remains of 156 mammoths, many of them young males, who perished between fourteen thousand and twelve thousand years ago,
during warming after the Ice Age.

Why did so many young mammoths perish here? The inexperi-
enced beasts may have ventured onto the river's spring ice, promptly
broken through, and drowned rapidly in the floodwaters beneath.
Their carcasses bumped downstream until they grounded, in the same
place year after year, an oxbow lake off a river bend. As the water receded, wolves and other scavengers feasted on the rotting beasts.

The bone accumulation resulted from the hazardous terrain, not
from mammoth choice or, for that matter, human activity. Vereshchagin

found a hunting camp of about the same period a few hundred feet
downstream, where the inhabitants lived off hares and grouse. Most
likely, they scavenged bones, and perhaps some meat, from the carcasses.

over a huge area of central and eastern Europe, from Austria to the shallow river valleys of Russia and the Ukraine. In the absence of natural
caves and rock shelters, the Cro-Magnons dug themselves into the landscape to escape the pervasive cold and frigid winds. Their food supplies,
their building materials, and other essentials, such as toolmaking stone,
came from sheltered river valleys, the open tundra, and mountain
foothills, distributed over an area that was only capable of supporting a
sparse population. Life during the run-up to the Last Glacial Maximum
required constant mobility and accurate intelligence about food supplies and raw materials far over the horizon.

All Gravettian groups maintained regular, if sporadic, connections
with neighbors near and far. They had to, to survive in a harsh world
where a chance hunting accident could wipe out all the men in an isolated band, or where death in childbirth could leave a group with no vi-
able future.[8] The ebb and flow of hunting groups must have begun early
in the spring, when river ice melted and snow cover thawed, during the
vital weeks when reindeer and other animals moved from winter pas-
tures. Small, temporary camps of hide tents appeared in river valleys and
on the tundra in a constant waltz of irregular movement determined by
experience and freely exchanged intelligence about bison, horses, reindeer, and other larger game. The popular image of Cro-Magnons generally revolves around their prowess with such prey, but much of their
subsistence came from smaller animals, whose meat yield per carcass
was surprisingly high. Like the Neanderthals and Aurignacians before
them, the Gravettians were opportunists, but with superior lightweight
weaponry, fiber nets (net fragments have been found at one site), and ingenious trapping technology that enabled them to harvest smaller game
whenever the opportunity arose . . .

Imagine a summer afternoon by a fast-moving stream, where dozens of arctic hares feed in the warm sunlight. At first light, the hunters have strung long fiber nets between two boulders at the entrance to a side gully close to the hares' burrows. Men, women, and children hide among the rocks, spears and clubs in hand. Several families have come together for the hunt. Over several hours, they encircle the unsuspecting animals, then move in, slowly waving sticks and reindeer hides. The hares circle and weave in confusion, clubbed down as they panic. Many stampede frantically for their burrows, only to end up thrashing wildly in the nets. Clubs pound and spears fly. Dozens of carcasses soon litter the ground, quickly piled up as the hunters cast around in a killing frenzy. As the sun sets, the people rapidly gut and skin the carcasses, carefully setting the pelts to one side and roasting some of the hares in the hot ashes. They'll dry the remaining meat; scrape, then peg out the skins; and preserve them for decorating clothes and making baby carriers . . .

With the approach of winter, everyone abandoned open camps and moved into small communities of more permanent, dome-shaped houses, where they may have remained for months at a time. They seem to have preferred areas of moderate relief, especially deep river valleys and some hilly terrain, for these were the landscapes that shaped their hunting strategies. As we shall see in chapter 10, some of these environments were remarkably productive for harvesting migrating animals.

Many bands returned habitually to the same locations over many generations. We know this because many major Gravettian sites, both in Moravia and far to the east, have yielded the oval foundations of many dwellings erected at the same place, creating a concentration of houses scattered over large areas. In some places, they settled in strategic locations used thousands of years earlier by Neanderthals. One such place was Predmost, which lies on a major game migration route at the southern entrance of the Moravian Gate, a natural defile to the tundra. Unfortunately, nineteenth- and twentieth-century limestone quarrying destroyed much of Predmost. Over a dozen archaeologists excavated at two long-used sites in advance of the quarrymen, but to little avail. Then a fire at the end of World War II destroyed all the artifacts and other finds. Thus, we have to rely on inadequate reports from an earlier era to tell Predmost's story.[9]

Thanks to recent excavations on an undisturbed spot, we know that Gravettians lived at Predmost about twenty-seven thousand years ago, in a time of briefly milder conditions when forests covered much of the landscape. Here they erected oval dwellings sunk partially into the ground, identified by dense concentrations of mammoth bones, some of which were carefully sorted. There are also traces of ritual observances, involving the careful burial of animals. At one spot, seven wolf graves were close to a large burned area surrounded by mammoth bones. In 1894, archaeologist K. J. Maska unearthed a large, elliptical human grave immediately west of the occupation area. Sharp-edged limestone boulders covered the pit. At least twenty individuals lay crouched in the 8.5-foot-(2.5-meter)-deep trench. Eight were adults; twelve were adolescents and children, the youngest only six months old. A fox skull was atop one of the skeletons, while bones from large and small animals abounded in the grave fill. The Predmost burial pit is the largest Cro-Magnon grave ever discovered. Alas, the find came before the days of modern fine-grained excavation, so we know nothing of the clothing and ornaments worn by the deceased, which might have revealed details of their social status.

As early as 31,000 years ago, and for at least 6,000 years, other Gravettians visited a location in Moravia known as Dolní Věstonice, overlooking the Dyje River.[10] The site lies in thick loess deposits that have long been exploited for brick making, a complex accumulation of Gravettian occupation that extends over a large area. Dozens of circular or oval dwellings once occupied terraces above the river, marked today by concentrations of mammoth bones, hearths, kilns for baking clay objects, and large quantities of stone artifacts and food remains. There are burials, too, including one remarkable triple grave of three young people dating to about 27,600 years ago: two men between sixteen and eighteen years old and a woman of about twenty. They lay side by side, their bodies strewn with red ocher. The three shared some unusual anatomical features, including impacted upper wisdom teeth, a very rare condition in the late Ice Age. They may have been related.

Dolní Věstonice is but one of many known Gravettian settlements in what was one of the more densely settled areas of central Europe twenty-five thousand years ago. Some of the best-preserved camps are far to the east, in the shallow river valleys of the East European Plain. Here people camped on well-drained river terraces, close to running water, in places where toolmaking stone and other raw materials were close to hand, often at junctions with small side valleys. The most famous eastern sites lie along Russia's Don River, notably the twenty-one archaeological sites of Kostenki. We described some of the discoveries at this location in chapter 6, for some of the earliest known Cro-Magnon communities lived here. The Kostenki 1 site lies near the mouth of a large side valley on a terrace above the Don. Here a series of occupations date to before about twenty-five thousand years ago, sites where people hunted predominantly horses and fur-bearing animals.[11]

Soviet archaeologist P. P. Efimenko carried out large-scale excavations at Kostenki 1 from 1931 to 1936. He exposed more than seventy-five hundred square feet (seven hundred square meters) in an area excavation that revealed fifteen large, debris-filled pits surrounding a line of central hearths. Efimenko believed that the Kostenki people lived in long houses covered with hides, a somewhat fanciful interpretation at best, but the same linear arrangement of hearths has turned up elsewhere. Another archaeologist, A. N. Rogachev, returned to Kostenki 1 during the 1970s and excavated another, smaller complex, this time an oval arrangement of debris-filled pits and three central hearths. These recall Dolni Vestonice and other such settlements to the west.

Another site, Avdeevo, near the modern Russian city of Kursk, dates to between 23,100 and 20,100 years ago. Once again, there are numerous hearths and debris-filled hollows. The surviving remains are reminiscent of late-prehistoric sites in the North American Arctic, where such features resulted from open-air encampments occupied briefly during periods when normally widely dispersed groups came together to trade, arrange marriages, settle disputes, and perform ceremonial observances.

Few excavations along the eastern rivers have given us much detail on actual houses. For specific information, we have to turn to a much later settlement, but one where the same general architectural tradition of semisubterranean dwellings persisted. The Mezhirich site, in the Dnieper River valley in Ukraine, dates to about 15,400 years ago, to a time long after the worst of the Last Glacial Maximum.[12] Winters were still intensely cold, though, so semisubterranean dwellings were essential in a landscape with little natural shelter. Mezhirich's inhabitants still built domelike oval dwellings, for there were no other realistic options (figure 9.2). They fabricated the outer retaining walls from patterned mammoth skulls, jaws, and long bones. Each house was about sixteen feet (five meters) across, partially excavated into the ground and then roofed with hides and sod. Nearby lay large storage pits filled with mammoth and other animal bones. The American archaeologist Olga Soffer has calculated that it would have taken about fourteen or fifteen workers ten days to construct all the Mezhirich houses. She estimates that three times the amount of labor went into an elaborate base camp like this one than into simpler ones.

Mezhirich and other large base camps yield large numbers of bones of beavers and other fur-bearing animals, taken during the winter months. Soffer believes that the people occupied winter base camps for about six months of the year, then moved out into summer encampments that were occupied for shorter periods of time, perhaps a month or so before the group moved on. Perhaps fifty to sixty people occupied each winter base camp, each house belonging to one or two families. For much of the winter they would have lived off food stored in permafrost

Figure 9.2 *Mezhirich reconstruction. Jack Unruh/National Geographic Image Collection.*

pits, with only occasional excursions to catch ptarmigan or to trap fur-bearing animals. Here and elsewhere, spring and early-summer reindeer migrations were a staple, a wide range of other animals, waterfowl, and sometimes fish. This hunting routine continued for thousands of years, from before the Last Glacial Maximum up to the end of the Ice Age.

EVERY BAND INVESTED enormous quantities of time in harvesting fur-bearing animals. They had to do so to survive long winters when fat was in short supply and people spent their days wrapped in furs to keep warm. Snares and traps were among the most important hunting weapons used by Cro-Magnons, for they allowed both men and women to catch a wide variety of smaller animals for feathers, meat, furs, and even delicate membranes and bird beaks. How do we know that they used such devices? Only indirectly, for all we find in Gravettian sites are the bones of the hares, foxes, and other animals that they trapped, then carefully skinned. Most snares, even larger traps, were made on-site, far from home, out of the most conveniently available materials, such as veg-etable fibers, saplings, and thongs. Their more durable components rarely, if ever, survive in archaeological sites, even if they may sometimes appear, unbeknownst to us, as enigmatic signs painted on cave walls.

We can only guess at the full range of Cro-Magnon trapping, by as-sessing the bones of their quarry that survive and by making intelligent and critical use of information from historic groups living in arctic and subarctic environments.[13] Both snares and traps are, for the most part, simple but highly effective devices that one can safely assume have changed little over the past twenty-five thousand years. Hares and rab-bits, for example, repeatedly follow the same trails, which show up clearly in snow. In historic arctic societies, the trappers would erect a light fence on either side of the trail, then set a snare in the trail attached to a pole balanced on an X-shaped pair of poles. The hare would enter the noose and trigger the trap, which would lift it, helpless, into the air. Such snares could be set in a few moments, were deadly efficient, and were surely used in Gravettian times. Judging from the large quantities of hare bones at some of their sites, they certainly were.

The Gravettians were masters of lightweight weaponry and the multipurpose artifact. Behind their weapons lay the Swiss Army Knife technology of blades and, increasingly, small bladelets, which formed part of what appear to have been more elaborate composite weapons with several parts. You can always identify a Gravettian site by the narrow, elongate backed blades, often called Gravette points, which were mounted in wooden spear shafts (figure 9.3). They were deadly weapons in skilled hands. Edge-wear studies tell us that the hunters also used such blades as knives, sometimes after they had been broken during the chase. Some groups also flaked small points with mounting tangs, used on lightweight spears. (A tang is the narrowed end of a point that is shaped to fit into the end of the spear shaft.) In central and eastern Europe, people used points that doubled as knives that have a distinctive shoulder, which may have made them easier to use.[14]

As time went on, the projectile points got smaller. Hunters relied more heavily on bone points with beveled bases manufactured from long bones broken into large splinters, then whittled and scraped into shape. The base bevel matched one cut into the head of the wooden spear shaft, with a step at the base of the bevel to prevent point and shaft from coming apart on impact. Some bevels had scratches to help them adhere to the shaft, the bond probably being "glued" with resin and then bound tightly with a thong. The result was a streamlined, effective weapon that could be used against large and small animals. Experiments by archaeologist Heidi Knecht with replicas show that such weapons had impressive penetrating power, sometimes even passing right through the carcass of the dead goat used in the experiments.[15] In southern Germany, some groups made large bone points from segments of mammoth ribs.

This efficient, light weaponry depended on the ability of the hunter to throw fast and accurately, as well as on spear designs that enabled him to change a broken head for a new one rapidly. If you look closely at cave paintings, you'll find many wounded animals, but very often no sign of the spear that hit them. This may be because the hunter used a composite weapon, with the lethal projectile point attached to a short bone or wood foreshaft hafted to the shaft. When the spear hit its prey,

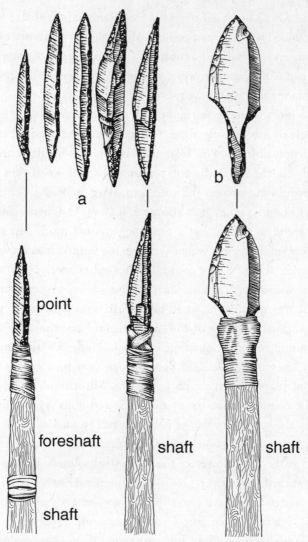

Figure 9.3 *Gravettian weaponry. (a) Finely backed blades, probably mounted as spear points, showing a direct mount and one using a foreshaft, which would break off when the spear hit its target, thereby allowing the hunt to mount a new point if need be. (b) A so-called Fort Robert point and its mount. (a) (b) Stone artifact drawings. Jean-Luc Piel-Desruisseaux, 2007. Outils Préhistoriques (Paris: Dunod). Reproduced with permission.*

the head would break off from the foreshaft, and the shaft would fall on the ground. The spearman could grab it, quickly fit another point to his weapon, and be back in action in a few moments. Composite spears combined with spear throwers, whose advantages we discussed in chapter 8, made for powerful weaponry in skilled hands.

Even more than the Aurignacians, the Gravettians were highly selective users of toolmaking stone. At Dolní Věstonice and other Moravian sites, people obtained flint from glacial outcrops in Silesia and southern Poland, among other distant places. Much of Kostenki's black flint came from outcrops on the southern margins of the Central Russian Upland, some 93 miles (150 kilometers) away. And there's another interesting trend in the lithics: a tendency toward much smaller artifacts made on small blades, which were then trimmed into crescent-shaped and backed forms. These could have served as spear-point barbs, lightweight weaponry that was ideal for people with spear throwers who were on the move. Some small blades are serrated, as if they were used for sawing. By the time of the Last Glacial Maximum, the Gravettians were hunting a very wide range of animals indeed with unmatched efficiency. They also possessed dogs, which may have been tamed thousands of years earlier; we do not know. Mitochondrial DNA suggests that the common ancestry of domesticated dogs lay among wolves in east Asia. The Goyet cave, in Belgium, has yielded what are claimed to be domesticated-dog bones in an Aurignacian level dating to as early as 31,700 years ago. However, the earliest absolutely certain dogs were found at the Eliseevichi 1 site, in the Dnieper River Valley, dating to between 13,000 and 17,000 years ago. They were somewhat like Siberian huskies, but with broad foreheads. Dogs would have been invaluable for driving smaller animals and for cornering them. They would also have been a source of food.[16]

For much of the year, Gravettian bands lived in isolation, in small camps of a few dwellings. Judging from the extensive clusters of dwellings at places like Kostenki 1 and Dolní Věstonice, several groups would come together during the warmer months, perhaps for some weeks or for shorter periods of time. If the density of house foundations is any guide, sometimes as many as a few hundred people would congregate at a specific location. As with historic societies that relied on hunting and foraging, these were occasions when the bands would exchange intelligence, settle disputes, and perhaps engage in communal hunting. This would also have been the time when marriages were arranged, exotic objects were passed from hand to hand, and cherished rituals unfolded . . .

The sun is low on the horizon at the end of the long summer's day. For once, the wind is calm, the temperature is warm, and insects are absent. The hide tents stand in a row along the river terrace, surrounded by huge piles of reindeer bones from years of spring and autumn hunts. Great fires flicker in the growing twilight. Women break bones for their rich marrow and thread reindeer meat on sticks. Children run between the tents; young men in shirts and light skin pants wrestle and tease one another. Some of them sit quietly and eye the girls, who are well aware of their scrutiny, knowing that marriages will be arranged in coming days. Near one of the hearths, four men sit around a reindeer hide on the ground. One of them comes from far downstream, from a land where water stretches to the horizon like the vast tundra. He scatters a handful of shells on the hide. The men lean over and the bargaining starts—over lumps of amber, wolf pendants, ivory bracelets, and other unfamiliar objects from afar.

As night falls, the bands feast on fresh reindeer meat. Then stories begin: tales of successful hunts, of great mammoths, of the deeds of ancestors and mythic hunters. A quiet old man with a weathered face starts a haunting chant about ancient times, about the Creator and the formation of the world and its animals. As the familiar story unfolds, a bone flute sounds, and people beat with their hands on taut skins. Young men and women start to dance in and out of the shadows. The dancing will continue all night, until dawn, in an ageless celebration of the continuity of life in a harsh world . . .

Such gatherings were the most important events in Gravettian life, which was lived in a vast landscape of long distances and large hunting territories. Despite the huge mileages, even widely separated bands kept in irregular touch, cementing relationships with gifts and the exchange of exotica of many kinds. We know of these exchanges from finds at sites like Kostenki, among them amber collected from deposits near Kiev, prized for its looks and magical properties. Mountain crystals at Mezhirich came from the south, from at least 62 miles (100 kilometers) away. Bands in the lower Dnieper Valley close to the Sea of Azov passed fossil marine shells from the Black Sea to the north. Some mollusks reached a settlement at Timonovka, on the Upper Desna River, more than 403 miles (650 kilometers) inland. Large numbers of Black Sea shells were found at Kostenki, more than 310 miles (500 kilometers) from the coast.

These exotics were far more than mere curiosities. They had important social meanings. A seashell, a piece of translucent amber, a rock crystal—each involved a transaction between two individuals, perhaps a token of friendship or the establishment of a bond that would endure, at a distance, for years, maybe even generations. Gifts and exchanges must have been the critical, and often near-invisible, links that helped cement relationships between small groups of people living in isolation from one another and often far over the horizon.

As always, we know nothing of the subtle rankings and hierarchies of Gravettian life, which must have been everyday factors in human existence. Only one spectacular discovery draws the curtain of history momentarily aside. The Sungir hunting camps, in northern Russia, lie in the Volga River drainage northeast of Moscow at fifty-six degrees north, at the time of occupation a mere 93 miles (150 kilometers) from the edge of the northern ice sheet. Mammoth and reindeer hunters visited Sungir repeatedly between about twenty-seven thousand and twenty-six thousand years ago, for the site then lay on a game migration route. Russian archaeologists are still deciphering the palimpsest of Cro-Magnon settlements here, including a remarkable cemetery.[17]

Nine burials have been found in Sungir's burial ground, the three best preserved being those of an older man, perhaps in his sixties, and two adolescents who were buried head-to-head. Each individual wore necklaces, bracelets, and brooches. Over thirteen thousand mammoth-ivory beads on strings once decorated the garments of the three most complete burials. The older man wore a beaded cap and perhaps a richly decorated leather tunic (figure 9.4). A red-painted schist pendant hung

Figure 9.4 *Reconstructed portrait of a man from Sungir and his garments and ornamentation. Painting by Liban Balák.*

at his neck. Twenty-four thin mammoth-ivory bracelets painted red and black lay on his forearms. The investment in these adornments was significant. Experiments have shown that a single bead would have taken more than an hour to fabricate. The beadwork on the older man's clothes would have taken about three thousand hours to make. The adolescents, who were apparently interred one thousand years later, wore five thousand hours' worth of beads, all individually drilled and carefully made to form interlocking patterns when strung, seemingly a much greater time investment. One male youth wore a belt of more than 250 arctic fox teeth. A schist pin at his throat may have secured a cloak. A heavy mammoth-ivory lance lay by his side, almost certainly a ceremonial artifact.

All the Sungir graves leave an impression of wealth and prestige, and certainly of elaborately decorated clothing. Such decoration may have reflected social status, kin affiliations, and even personal skills. If the lavish adornment of the two adolescents is any guide, marks of social distinction at Sungir passed from one generation to the next. Without question, Cro-Magnon society was far more colorful and complex in its social relationships and spiritual beliefs than we assume.

THE ARCHAEOLOGICAL FINDS at places like Dolní Věstonice and Kostenki are only the tip of a cultural iceberg. So are the Gravettian ornaments found there: perforated carnivore teeth, shells, and fragments of worked bone, ivory, and clay used as pendants or beads. Beyond numerous bone and ivory fragments engraved with geometric lines, and some stylized animal and human figures, there is none of the rich animal imagery found in western caves such as Chauvet. Here the canvas was bone and ivory, the art on a more diminutive scale. Nothing remains of the rituals and spiritual beliefs that lay behind Gravettian societies, just occasionally a few more-elaborate ornaments, like the mammoth bracelets worn by the man at Sungir and what may be ivory headbands ornamented with geometric patterns found at Avdeevo and Kostenki. At the Pavlov site in Moravia, artists carved in ivory, including a series of finely carved rings, as well as possible headbands carved with curved and arched lines.

The artistic conventions are remarkably standardized, something that is also true within historic Arctic art traditions, again a reflection of long distances and sparse populations that kept in touch with one another. For the Gravettians, part of this seeming uniformity is a constant artistic presence: voluptuous female figurines. So curvaceous are they that early archaeologists named them Venuses, an inappropriate but convenient term that haunts the figures to this day.

Venuses occur all across Europe, from Ukraine to southwestern France. No one knows where the convention originated, but it may have been in central Europe, where artists fashioned the figures in ivory and baked clay, the earliest known ceramic objects. The earliest known example, from southern Germany, is about thirty-five thousand years old.[18] The so-called Black Venus, from Dolní Věstonice, stands with her legs together, belly button clearly marked, arms not shown. Her hips are broad and fat; her pendulous breasts hang down her torso (figure 9.5). She has no features except for a vague slit for her mouth, and she has no hair. Another celebrated Venus, from the Willendorf site, in Austria, estimated to be about twenty-nine thousand years old, is in the same style (figure 9.5). Again, the head is featureless. Crisscrossed lines represent her hair. Her breasts hang large. Her diminutive arms rest atop them. Willendorf's hips are enormous, so big that many researchers wonder whether she was steatopygous, a condition that produces a large fatty accumulation around the hips and thighs, a characteristic of some San women in Africa.

To the east, at Kostenki 1 and other sites, artists used ivory and clay to create the figurines, again with pendulous breasts and exaggerated hips. One mammoth-ivory figure from Kostenki 1 came from a pit by the communal central area. She appears to be wearing some form of cap and maybe a belt. This figure has well-muscled legs. Her toes are pointed at each other. The eastern sites also yield abstract representations of women and of breasts. However, it is the placement of the Venuses in the eastern sites that is most fascinating. One figure stood upright, leaning against the wall of a small, hourglass-shaped pit facing a central row of fireplaces. She stood atop a layer of silt, alongside burned ivory fragments, some bone plaques, and stone artifacts. The pit had been filled with fine silt and red ocher, then covered with a mammoth shoulder blade. At Avdeevo, about 78 miles (125 kilometers) away, some twenty-five female figurines were found in pits surrounding a central row of fireplaces and depressions. One pit contained three clusters of artifacts, including two statuettes arranged head to toe and back to front. They interlock perfectly and were made from the same piece of

Figure 9.5 *Venus figurines. (left) Stylized female figurine from Dolní Věstonice. Werner Forman/Art Resource, NY. (right) Willendorf, Austria. Erich Lessing/Art Resource, NY.*

ivory. Clean river sand filled the bottom of the pit, followed by a dark layer that contained a lion skull and a final layer with occupation debris and the remains of dozens of wolverine skulls and jaws. At another site, Gagarino, the inhabitants dug ten niches into the loess, equidistant from one another. Each contained a voluptuous female figurine, one of them incomplete. Perhaps this was a birthing house, or the dwelling where the sculptor lived. More statuettes were found in a pit adjacent to the house, together with the foot bones of foxes, as if the figures had once been wrapped in fox pelts.

The same artistic tradition spread to the west, to caves in southwestern France, but for some reason not into northern Spain. Cro-Magnon groups visited the Laussel rock shelter, near Marquay in the Dordogne, for thousands of years, but it was a Gravettian artist who pecked a Venus figure into the wall. The woman wears some form of ill-defined headdress and looks to the right at her right hand, which lifts a bison horn adorned with thirteen slashes. Her left hand rests on her swollen belly. With her heavy breasts, massive hips, and pubic triangle, the Laussel Venus follows the usual conventions. She was merely sculpted on a limestone block rather than as a stand-alone figure. The "woman with a horn" is the best known of a series of human figures from Laussel, also including one of two people, another of a woman with a gridded headdress, and a third of a standing male figure seen from the right, apparently wearing a belt and pants (figure 9.6).

Perhaps the most famous French figure is an ivory head of a woman from La Grotte du Pape, at Brassempouy (figure 9.6). Known as the "hooded woman," she has long hair that cascades down her neck, and an almost impressionist execution of her face makes the figure change appearance from various angles and in different lights. The Brassempouy finds define the artistic tradition behind all these Gravettian figures: the artists executed only the essential features, leaving much to the imagination. The curves and swellings of the figures evoke female anatomy, even reproductive qualities.

Many historical tourists have argued over the meaning of these remarkable figurines in what the paleontologist R. Dale Guthrie rather deliciously calls "a comfortable fuddle."[19] Inevitably, there are those

Figure 9.6 *The Brassempouy Venus (right). Réunion des Musées Nationaux/Art Resource, NY. Venus bas-relief from Laussel (left). Erich Lessing/Art Resource, NY.*

who believe the figurines are associated with a universal Mother Goddess cult, which, at best, seems far-fetched. The figures' prominent anatomical features hint at some association with birth and fertility cults, but common artistic conventions can be said to be just that, nothing more. In most living hunter-gatherer societies, the people are well aware of the severe limitations placed on a band by too many children, who require time-consuming investment before they become productive members of society. For instance, the anthropologist Richard Lee found that the !Kung (! is a palatal click) San of the Kalahari Desert have stringent sex taboos during the first years of a baby's life.[20] Living as they did in harsh environments where they depended on unpredictable animals, the Cro-Magnons must have been well aware of the dangers of undue fertility. For this reason, whether the Venuses were symbols of fertility, birth, and renewal is very much an open question.

We tend to skirt around sex when talking about the Cro-Magnons, who were inveterate sketchers of humans, often in what can only be called sexual poses. Guthrie argues that the Cro-Magnons had sexual feelings, just like modern people. He points out that where there is enough detail to be sure, females are always depicted without clothes. Almost invariably, too, they are women in their prime reproductive years, with round figures from about age seventeen to forty. He also observes that most of the women have full figures and are even corpulent, characteristics that both biologically and socially represent fecundity. In what may be overtones of sexual invitation, the artists stressed the erotic parts of the body—breasts, buttocks, hips, and thighs—whereas the arms, faces, hands, and even ears and eyes are de-emphasized. In the same vein, there are also hundreds of explicit human images on the walls of Cro-Magnon caves, some little more than graffiti, others depicting sex organs, a few copulation. Guthrie writes, "So these distant ancestors, in good humor, made marks of passion and desire in ivory and on limestone walls. We have these identical traits of overwhelming obsession. They are not the refuse of illicit orgies nor are they the accoutrements of holy shrines, just casual breasts and vulvae scattered among lines of tail and antler— marks that played with the brain and made life more interesting."[21]

If Guthrie is on the right track, then the Gravettian Venuses, with their fat-filled breasts and buttocks, were an erotic signal of attractiveness in a society where fat consumption and increased weight helped survival. The figures are depictions not of pregnant women, but of full-figured women carrying fat in ways that are very different from pregnancy, when women carry their young high. Also, the proportions of the fat are displayed in ways that are consistent with obesity. So the voluptuous women of Cro-Magnon art may symbolize erotic potency, the periods of optimum sexual desirability when they carried weight: after the growth spurt of puberty and when babies had been weaned. Some of the doodles may just reflect teenage hormones operating at full bore.

There's an unexpected logic in Guthrie's theory, and it jibes well with the essentially simple foundations of Cro-Magnon life, which revolved around the realities of periods of abundance and scarcity. Above all, their survival depended on efficient harvesting of the bestiary that surrounded them on every side.

CHAPTER 10

The Power of the Hunt

THE VÉZÈRE VALLEY, FRANCE. The men have watched for reindeer for days. They sit, warmly clad, on high crags above the valley in the cool fall sunshine of twenty-five thousand years ago. Below them, the mud brown Vézère flows sluggishly, riffling gently across the shallows of the ancient crossing place. Nearby, the camp is abuzz with expectation, spears and spear throwers at the ready, scrapers and knives close to hand. A still-warm breeze stirs the leaves of the dark conifers at water's edge. And the hunters wait and wait.

Next day, a gray morning dawns. Light mist whispers atop the trees. Croaking ptarmigan feed in the short, still-green grass of autumn. There's a soft backdrop of movement in the gloom, as if of gentle breathing. Suddenly, a cry summons the hunt. The reindeer are coming! Excited men, women, and children leap to their feet and grab weapons and tools, hunters to the fore, lining the riverbanks, nestled among boulders and trees. Each hunter carries a bundle of spears and his spear thrower. The press of reindeer arrives, heading northward, a crowding host of beasts fifty yards wide. The stream of animals parts on either side of a grove of trees and then comes together again as the host approaches the sloping riverbank. Moments later, the leaders enter the ford and splash across to the other side. Those behind them move inexorably in their footsteps, an orderly, intent highway of living things. Some cross in deeper water with powerful, buoyant legs, antlered heads atilt, packed so closely together that you could cross the river on their backs.

Dozens of beasts flow to the other bank before the hunters strike.

They rise from behind rocks and trees and move forward, aiming at the nearest animals in the shallows. The thud of spears hitting living flesh echoes across the valley. Crossing reindeer fall before the onslaught, trapped in the river by the momentum of the animals pressing forward and cutting off their retreat. Those approaching the ford break forma-tion. They run back and forth in short bursts, upstream, then down-stream, before stampeding along the riverbank and out of sight. The ford is a battlefield of dead and wounded beasts. Wounded reindeer flail helplessly, trying to rise on their haunches and escape. Others raise their heads and look at the hunters as they die. The men prance among them, jumping nimbly clear of slashing antlers. They deliver death blows with spears and clubs. Some they kill with sharp stone knives, severing their spinal columns with a quick slash. Reindeer blood pumps into the stream. The Vézère turns dark red. Long crimson streamers of brighter color streak far downstream before the current disperses them. Minutes after the hunt, more than thirty reindeer lie still at the ford.

Now the hard work begins. The men drag the carcasses from the wa-ter to the sloping riverbank. There they split open the bellies with a sin-gle long stroke, reaching inside the stomach cavity to disembowel their prey. They remove the liver and kidneys, then detach the hindquarters from the trunk. Reindeer tongues are a delicacy, so they avulse them by slicing through the skin under the tongue. People grab hunks of fresh meat and fat, wolfing them down as the blood spills over their chins and clothes. Meanwhile, deftly and without fuss, men and women skin each animal, helped by their sons and daughters, who learn the art alongside them.

Back in camp, men and women dump heavy loads of skins and flesh. While the hunters dismember the carcasses, cut off the flesh, and hang it up in strips to dry, the women scrape the fat off the skins and then peg them out on frames or the ground to dry. People reach for a limb bone, smash it open, tap it on a rock, and suck out the yellow marrow. Mar-row is fat and is always a delicacy. Reindeer tongues cook on sticks over the hearths. The bands will eat well tonight.

As the shadows lengthen, the men sit around a fire just outside the camp telling tales of the hunt. Their sons sit with them, hanging on every word, reveling in the storytelling. The gentle sounds of animal movement are still a backdrop to their consciousness, as the endless migration ebbs for the night. A man suddenly points. Another reindeer herd is crossing the ford just upstream of the camp. This time, the hunters only watch. The beasts cross the stream in a solid mass, then move onto higher ground. They head straight for the campfire, then split into two groups as they flow past the hearth, completely oblivious to the sitting Cro-Magnons. Three of the beasts come so close that the men wave and shout at them with an easy familiarity and literally push them away. Moments later, they are gone, melting into the dark shadows. The people around the fire are completely unfazed, for they think of the reindeer as friends, living beings just like themselves.[1]

Day after day, the hunters prey on the migrating reindeer as they cross the river. Ambush after ambush does nothing to stem the purposeful herds. Then, after a week or two, the seasonal migration ends abruptly, almost as if a faucet has been turned off. The hunters dry the last of the flesh and stake out the final hides. Everyone gorges on fat and marrow while they can. The women have smashed up spongy backbone and limb ends and boiled them in simple containers to extract as much grease as possible. They store as much reindeer fat as they can in leather bags, to be hung high off the ground in their rock shelter bases for the winter. The most important hunts of the year have ended, but memories of them will endure throughout the approaching cold season . . .

RANGIFER TARANDUS, THE reindeer, must have been the stuff of legend and ritual to the men and women who lived off them. Thousands of fragmentary *Rangifer* bones lie in the occupation levels of Gravettian rock shelters in southwestern France, but they presumably represent but a fraction of the enormous numbers of reindeer taken by Cro-Magnon hunters every year. Many of the bones in the shelters may have come from solitary hunts during summer and winter, when small groups of reindeer browsed in quiet river valleys. A hunter and his son might stalk a beast, or a group of young men go after an animal crossing a stream, just as they did with other game, like ibex and red deer. But almost all the reindeer taken during the year must have come from the harvests of spring—when thousands of them would have migrated to the foothills of the Pyrenees to avoid the heat and mosquitoes in the deep river valleys of the Dordogne, among them the Célé and the Vézère—and fall, when they would have returned from higher elevations (figure 10.1).

I use the word *harvest* because the spring and fall hunts were just that: organized reapings involving several bands and considerable numbers of people engaged in mass kills and large-scale butchery. Why, then, do we find relatively few traces of the enormous numbers of carcasses butchered and processed by the Gravettians? The answer may be a simple one. Most of the reindeer bones we know about come from caves and rock shelters where people lived. Most spring and fall hunts were probably stayed from temporary camps. By the time the hunts were over, huge piles of rotting reindeer carcasses would have surrounded the encampments. The fresh meat would have attracted lions,

Figure 10.1 *Reindeer migration near Abri Pataud in the Vezère Valley in Gravettian times, c. 25,000 years ago. Illustration by Eric Guerrier. Musée de l'Abri Pataud aux Eyzies-de Tayac, Dordogne and Institut de Paléontologie Humaine. Drawing executed under the supervision of Professor Henry de Lumley.*

hyenas, and wolves like magnets. Even upwind, the smell of decaying carcasses, and the flies, would have been pervasive. No sane hunting band would have lingered at such places longer than necessary, even with large, smoky fires burning. They would have collapsed their tents and moved on, carrying large amounts of meat and hides back to their base camp to be dried, smoked, and processed for later use. Grease rendering by boiling softer body parts would have yielded large quantities of fat that would keep for months. Within a few years, the piles of reindeer bones at the temporary camps would have decayed and vanished, leaving nothing for the archaeologist to explore thousands of years later. Some groups did, of course, prey on migrating herds from rock shelters, like the Abri Pataud, near the Vézère, described below. But the lack of known kill sites is in dramatic contrast to the prevalence of earlier kill sites in the Don Valley, mentioned in chapter 6.

The people would have hunted migrating reindeer in spring, but like humans, the beasts would have been lean and fat deprived after the long winter months. The herds were in their best condition in late summer and fall. Their coats were soft and springy. A mature reindeer can carry 30 pounds (13.6 kilograms) of fat under its hide. This is why the fall hunt was of prime importance to the Cro-Magnons, for it was then that they acquired their stocks of fat, hides, and dried flesh to meet winter needs. We know just how important it was, for little has changed for subsistence hunters since then. As was often the case for historic caribou hunters in Canada's Barren Lands, there must have been years when there was insufficient fat to provide both food and fuel. Yet the people managed to stay alive in unheated dwellings because of efficient clothing, by spending a great deal of time under heavy furs, by sleeping in close proximity, and by burning off the fat stored in their bodies from summer. For about six to eight months a year, depending on the severity of the winter, most Cro-Magnon bands would have relied on stored food from late-summer and fall hunts. Spring and early summer would have been a stressful time of potential hunger even in good years.

Rangifer tarandus is an eclectic feeder, consuming lichens and other foods in delicate plant communities. Reindeer move constantly, which minimizes the effects of fluctuations on their food supplies. They also dislike both heat and mosquitoes. Their seasonal migrations were the Cro-Magnon equivalent of the annual inundation that watered the fields of ancient Egyptian farmers thousands of years later. Not that the migrations were a certainty: quite the contrary. Like the Nile, which is capricious in its flooding, reindeer could behave unpredictably. They would change migration routes without apparent reason, use several unfamiliar defiles, or arrive in far smaller numbers than usual. But if the bones from Vézère rock shelters are to be believed, they were among the familiar staples of Cro-Magnon life for thousands of years.

Reindeer teeth are a fascinating chronicle of Cro-Magnon hunting life. The Canadian archaeologist Bryan Gordon has studied the complex movements of late Ice Age hunters in southwestern France by measuring the growth increments on reindeer and caribou teeth.[2] During the warm months of summer, the teeth display thick, clear growth marks, whereas those for winter are dark and thin—somewhat like the familiar growth rings displayed by tree trunks. Gordon compared ancient teeth with those from modern beasts, which gave an indication of the ages of reindeer killed at different Cro-Magnon sites. He then matched the teeth of two-year-old reindeer from La Madeleine rock shelter, on the Vézère, with those of two-year-old reindeer from Canecaude, in the Pyrenees, 124 miles (200 kilometers) away. All the teeth had two winter increments, but those from La Madeleine had thin spring ones, whereas those from Canecaude bore increments characteristic of late-spring and fall growth. The two sites are sufficiently far apart to represent the normal positions of winter and summer ranges of moving reindeer herds. Gordon believes that there were spring and fall migrations that had reindeer herds traveling between 180 and 250 miles (200 and 400 kilometers) in a few weeks—an equivalent distance to that traveled by caribou in the Canadian Barren Lands.

Gordon's research hints at very complex group movements that persisted over thousands of years. By the closing millennia of the Last Glacial Maximum, after eighteen thousand years ago, there were eight reindeer ranges in the West that we can identify, three in southwestern France, and others to the north and east. Both before and after the Late Glacial Maximum, the annual round of Cro-Magnon groups in the

West revolved around these ranges and the antlers, fat, hides, and meat they provided.

THREE GREAT CRO-MAGNON rock shelters lie near Les Eyzies: La Ferrassie, occupied by both Neanderthals and Cro-Magnons; the Abri Pataud, predominantly a Gravettian camp; and Laugerie Haute, most intensively used somewhat later. The inhabitants of all three preyed on reindeer, red deer, and, of course, other game. They used these shelters as base camps over many thousands of years. By about twenty-five thousand years ago, La Ferrassie's overhang was almost completely shattered by ice and frost, rendering it virtually useless as a stopping place. By then, Gravettian groups had long occupied the Abri Pataud (the rock shelter of Pataud), at the foot of the cliff on the other side of the Vézère River.

During the nineteenth century, the Pataud family farm nestled under the cliff, complete with bread oven, pigsties, and toolsheds. Farmer Martial Pataud found the prehistoric shelter while constructing a track to his buildings, but had absolutely no interest in archaeology. With Pataud's reluctant permission, a stream of early archaeologists dug small trenches into the deposits under the farm. Fortunately, he resolutely opposed any large-scale digging on his property. There matters lay until 1949, when the Harvard archaeologist Hallam Movius rented a tract of the property for a trial dig. The results were so promising that he purchased the land encompassing the shelter in 1957 and deeded it to the French National Museum of Natural History. Movius demolished the farm's barn and used the stone to build a new house for Martial Pataud's descendants a short distance away. He then directed six seasons of digging into the depths of the Abri Pataud, from 1958 to 1964. The old farmhouse became the site laboratory.

The Abri Pataud dig was a state-of-the-art field investigation, conducted with all the rigor of mid-twentieth-century science (figure 10.2).[3] This was no quick probing of densely backed occupation levels, as had been the case with earlier diggings at places like Laugerie Haute, but a minute dissection of a rock shelter occupied first by some Aurignacians and then mainly by Gravettians from about 34,500 to 20,500 years ago.

Zooarchaeology, or "the bones come together, bone to its bone"

How do we know about Cro-Magnon hunting practices? Our information comes from two sources. Modern-day accounts of traditional hunting in the Arctic are a mine of information, given the reasonable assumption that there are only a limited number of ways in which you can drive reindeer or trap an arctic fox. The bones of the prey butchered by the hunters and their families are the most valuable archive, but one that is frustratingly difficult to piece together. Animals were not only meat and hides; they provided raw materials for antler and bone tools, bags, thongs, and clothing. Tongues and entrails were eagerly devoured; long bones provided marrow; animal fat gave energy and warmth throughout the year. Usually, all that remains are discarded, shattered bone fragments, trodden into rock shelter floors and thrown into hearths or pits. When push comes to shove, almost all animal bones from archaeological sites are too fragmentary for identification. Fortunately, however, some body parts, such as teeth, jaws, horns, and some limb bone joints, are readily identifiable when compared with either modern-day animal bones or well-preserved fossil skeletons. You can easily distinguish the bones of *Rangifer tarandus* from those of a wild horse, for example.

The plot thickens once the identifications are made, for it is not just a matter of totting up the percentages of the bones. What one researcher once called the "archaeological animal" (the counted bones) is a very different beast from the actual animal that was killed, butchered, and carried, at least in part, back to camp. By carefully inventorying the individual body parts in an occupation level, one can sometimes obtain a count of the minimum number of animals in the collection, but it's an approximation at best. Once you have a count of individuals, you then confront other questions. Are the changes in the numbers of, say, reindeer and horses found in, say, an Aurignacian occupation layer and a Gravettian one a reflection of changed hunting practices, or simply chance? Does the overwhelming dominance of reindeer in the layers of the Abri Pataud mean that the inhabitants pursued them to the exclusion of other beasts? Were these chance hunts, or the result of harvesting spring and fall migrations? Some answers can come from identifying the ages of the animals, by examining horn cores or upper and lower jaws with at least some of the teeth intact. Growth rings in reindeer teeth are helpful for studying ages and migration seasons, too. A high proportion of juveniles or young adults can imply selective hunting, as can a focus on older beasts. The Cro-Magnons were such efficient hunters that they took prey of all ages.

A great deal of zooarchaeology involves turning percentages and animal counts into meaningful interpretations of human behavior. One approach involves studies of living hunters, notably the Nunamiut caribou hunters of Alaska, studied by archaeologist Lewis Binford some years ago. He found that Nunamiut food-procurement strategy was based on complicated decisions that involved not only the distribution of food in different seasons but also the storage potential of different animals and their parts, as well as the logistics of obtaining, carrying, and storing meat. Was it, for example, easier to move people to the herds or to carry meat back to base? We have to remember that fragmentary animal bones in the laboratory reflect intangible but logical decisions made thousands of years ago.

Every Cro-Magnon man or woman knew how to butcher animals of every size with sharp-edged stone knives. They acquired the skills while still children, perhaps first with reindeer and other medium-sized animals that were relatively straightforward to skin and dismember. Arctic foxes and other fur-bearing animals were another matter, for the pelt had to remain as intact as possible. Thousands of fox bones from Cro-Magnon sites show how the skinner cut through the hide at the feet and worked up from there. This kind of skinning was very different from the near-industrial-scale butchery needed when harvesting reindeer or horse migrations. Studies of breakage patterns show us how the Cro-Magnons disarticulated limbs, broke skulls open to get at tongues, and smashed limb bones for the marrow, using simple, effective methods that varied little over the centuries and are still in use today.

Like so much else in Cro-Magnon archaeology, their food remains are a palimpsest of information, teased out from seemingly insignificant bone fragments that usually reflect a few seconds of human behavior thousands of years ago.

It's safe to say that this remarkable excavation set new standards in rock shelter research that have been the basis of all later Cro-Magnon investigations. Over seven seasons of slow-moving digging, the excavators recovered almost 1.5 million finds—artifacts, food remains, and art objects—in packed occupation layers compressed under huge boulders from the shelter roof that had to be removed before excavations could commence.

Excavating the tightly compacted layers of a rock shelter is about the most challenging form of archaeological excavation of all, especially if you are as meticulous as Hallam Movius, who was obsessed with accurate recording and comprehensive data recovery. He installed a pipe grid over the archaeological deposits so that he could measure the exact position of every significant artifact, every feature. (He excavated before the computer era came to archaeological digs. Today's archaeologists rely almost entirely on electronic recording.) Stone-tool making debris, animal-bone fragments, hearths, and clusters of tools used for different activities chronicled dozens of short visits to the rock shelter. Using brushes, dental picks, and trowels, dozens of archaeology students removed each occupation layer inch by inch, spending as much time recording as they did excavating. The excavators plotted every hearth, as well as the artifacts and food remains around it, teasing details of the activities carried out there from seemingly inconspicuous objects. Just maintaining the diagrams of the stratified natural layers and various human occupations was a full-time job, as was plotting the emerging levels. The bags of finds went to the field laboratory, where Movius and

Figure 10.2 *Abri Pataud excavations. The metal pipe grid was used for measuring the positions of individual finds. © President and Fellows of Harvard College, Peabody Museum #2004.24.33363.*

more students sorted the material. They monitored the artifacts carefully for significant changes or innovations, worked on the dig, and sorted finds. The laboratory research took years. Almost fifty thousand artifacts came from about forty encampments at the Abri Pataud, grouped in fourteen archaeological layers. The burins alone formed an important doctoral dissertation that took years to complete.

Studies of the collections continue to this day, but the general history of the Abri Pataud is now well known. Some Aurignacians were the first visitors to the shelter, between about 34,500 and 29,000 years ago. They left their split-based bone points, burins, scrapers, and other distinctive tools behind them. Between 28,000 and 20,500 years ago, Gravettian visitors used the shelter intensively in a series of occupations that are divided archaeologically into four stages marked by different tool types. From our point of view, the differences are minor and need not detain us here. Every band that used the site made burins and scrapers, used for processing antler, bone, and skins. They also manufactured enormous numbers of sharp Gravettian knives or projectile points that grew progressively smaller as time went on, a reflection of lighter, more efficient weapons, some of them armed with barbs as well as single points.

One Gravettian occupation, dating to about twenty-four thousand years ago, centered on a substantial, tentlike structure erected between the cliff and some boulders at the front of the shelter (see color plate 5). Poles sloping between the floor and the back wall provided a sturdy framework for a covering of sewn hides. Piles of stoneworking debris lay outside the dwelling. You can imagine the great rock shelter on an unusually warm late-winter day. The sun casts long shadows across a litter of much-fragmented bone and stone blades, interspersed with boulders. Spears lean against the back wall, where leather bags of frozen reindeer fat hang high above the floor. A fire of dry branches sends white smoke to hover under the overhang on this windless morning. A woman in a thin shirt and long pants tends the fire, which lies outside a weathered hide shelter erected around stout poles. Inside, an old woman swathed in reindeer hides sleeps restlessly next to a smaller fire. The smoke rises between rows of socks and boots hung up to dry. Two men in summer parkas sit on boulders outside the tent. The younger one strikes blades off a flint core. His father uses a burin to groove fresh bone, soaked overnight in water to soften it. The movement of the burin fills the rack shelter with scraping noises, which mingle with the shouts of children playing on the slope down to the frozen river below. A pervasive scent of human sweat, rotting flesh, and drying furs drifts across the camp . . .

The packed layers of the Abri Pataud tell us that Gravettian life by the Vézère changed little over a span of more than seven thousand years. The inhabitants hunted wild horses—compact, well-muscled beasts with small heads—and aurochs, whose bones appear in the shelter. Above all, however, they preyed on reindeer, whose bones represent between 80 and 90 percent of all the animal bones in the occupation layers. Much of the time, the climate was cold and dry, the valley covered with patches of short grass, with coniferous trees on hillsides and near the rivers. During brief warmer spells, the hunters targeted more stags and wild boar. Most of the time, though, the staple was *Rangifer tarandus*. They also hunted many smaller species, many of them fur-bearing animals, including rabbits, and snared ptarmigan, grouse, and ducks. By any measure, these people were efficient, ingenious hunters. Everyone, man or woman, adolescent or small child, lived in intimate association with their surroundings in a way that is unimaginable to modern West-erners. Above all, their art tells us that they were consummate observers.

THE CRO-MAGNONS SPENT most of their time not hunting but watching—for hours, days, even weeks. They lived among a bestiary whose behavior they knew as well as their own. They were familiar with the animals routines of feeding and drinking, birth, courting, and death. Every hunter knew when reindeer were in top condition, the subtle color changes in the fur of arctic foxes, the sometimes dramatic changes in appearance that accompanied different social behaviors in all manner of beasts. Every Cro-Magnon knew more about animal behavior than most twenty-first-century biologists. They lived so close to their prey that they were part of their quarry's lives. It was this knowledge, based on meticulous observation, that made them successful hunters.[4]

You need only look at the details of cave paintings and engravings on antler fragments and weapons to realize this. Individual paintings and smaller engraved pieces tell us much about late Ice Age animals.[5] A reindeer raises itself to a standing position by using its forelimbs. A male sniffs the genitals of a female as they lie engraved on an antler beam. The differing sizes of their antlers portray their sexes. Horse manes are full, just as they were in life; the bison depicted on cave walls lack the thick pads of hair characteristic of American bison, an adaptation to their violent head-to-head clashing.

Cro-Magnon art, when examined judiciously, is a mine of information about the species the people hunted. For instance, a hunter watching large herbivores spends at least half of his time watching them eat. Cro-Magnon artists sometimes depicted habitual grazers doing just that. Horses normally prefer medium-height grasses but occasionally will stretch their necks downward to crop much shorter grass. The artists painted or engraved some of them feeding in this manner, as they did such quiet activities as a bison grooming its flank. But they drew many more animals with their tails alert or in threat postures. The artists had a "distinct taste for excitement."[6] At La Mouthe cave, near Les Eyzies, we see two mammoths sparring head-to-head, as they also do at other caves, like the large cavern of Rouffignac. There are courting scenes between reindeer. The artists depicted *Rangifer* in autumn, with fully developed antlers, and the distinctive summer and winter coats of horses, which in very cold environments, in winter, grew hair around the jaw that looked almost like a beard. All the horses in Grotte de Chauvet wear such winter coats. Bison painted in northern Spanish caves like Altamira often have reddish bay hair and black legs, manes, and tails. Those further north bear less-spectacular coats, often gray-brown.

The Cro-Magnons surrounded themselves with depictions of the most challenging and formidable members of the bestiary, some of whom were too dangerous to be staple prey. It's a mistake to think of Cro-Magnon art as being confined to caves and other places far from daylight. Every hunter carried a small arsenal of more-permanent weapons (as opposed to disposable projectile points) that he decorated for himself, perhaps with slashed lines or geometric or curved patterns, or with depictions of the animals that surrounded him on every side. There are doodles and crude sketches, simple engravings, and the occasional depictions that are masterpieces, executed by individuals with true artistic talent. Only a small fraction of this mobile art has survived the millennia. More of it must have adorned perishable substances like hides and wood than more-durable and harder-to-work antler, bone, clay, ivory, and stone.

The archive of drab stone tools and dark bones that has come down to us does a disservice to its creators, and to what must have been bright and colorful societies. Body paint denoted social status and affiliation; bright and muted colors alike marked spears and other artifacts; bird feathers may have formed headdresses. People may have even made brightly painted wooden masks. Their taste for color and contrast was surely as well developed as our own. Painting and engraving were ubiquitous in Gravettian life, as much a part of the fabric of daily existence as spears or knives. Much of the art was pragmatic and personal, part of a colorful and dangerous existence, a way in which people marked their possessions, commemorated their social affiliations, and expressed their individuality.

Some of the wall art may also have been purely decorative. At the

Roc-de-Sers rock shelter, in France's Charente, a frieze of twenty engraved rocks once ran the length of the shelter wall. One of the boulders at the mouth of the shelter bears an image of two fighting mountain goats. They lock horns, as ibex do during the rutting season as they win mates. At another shelter, Le Fourneau de Diables, near Les Eyzies, artists carved two aurochs on a large rock, which once stood at a slight angle, its edges buried in scree. The great animals, with their lyre-shaped horns, were but two of twelve images engraved and carved on this one rock. Such friezes were visible in daylight and in common view.[7]

In the profound darkness of subterranean chambers, it was a different matter. Here the powerful forces of the supernatural realm pressed on visitors, apparently from behind the rock faces. Paintings flickered in an illusion of movement in the light of a pine brand or a fat lamp. One can imagine visitors of all ages clambering into the depths of the Pech Merle cave, near Cabrerets, in the Lot. The large cavern boasts lower chambers that were never painted, but near the ancient, now-blocked entrance a jumble of finger flutings appear on the ceiling, a large area of rectilinear and curved lines, which include some naturalistic images, among them a mammoth. Under the floor below it lies a woman facing left, her body bent forward, breasts hanging, large buttocks clearly shown. She may have worn a headdress. She looks very similar to an engraved female figure at the Cussac cave, in the Dordogne, dated to about 25,100 years ago. Pech Merle also boasts one of the most famous of all Cro-Magnon friezes. Two black horses face in opposite directions, the head of the right-hand animal emphasized by the natural shape of the rock (see color plate 8). Large black dots surround the beasts and adorn their bodies. Six black hand stencils lie close by, while a red fish appears above the horses. The right-hand horse was painted 24,640 years ago.[8]

The meaning of the cave art eludes us, except, perhaps, for the hand imprints. Hand imprints and animals—the association goes back deep into Cro-Magnon history to the Aurignacians and Grotte de Chauvet with a persistence that suggests that people acquired some form of power from contact with dark rock faces beneath the earth (see color plate 9). Powerful testimony also comes from the Gargas cave, in the Pyrenees foothills.[9] Generations of visitors—men, women, children, even infants—left their hand impressions on the walls of the cave's lower level. Some of them lie close to cracks in the walls filled with bone slivers that were deliberately placed there. One of the bone offerings dates to 26,860 years ago. There are more than two hundred hand stencils in this one chamber alone. A small niche elsewhere in the cave frames a black hand stencil surrounded by the relief of the wall. All the fingers are shortened, as they are in many Gargas handprints, presumably deliberately, for reasons that elude us. The participants used red iron oxide or black manganese oxide to outline their hands, which created the impression that their hands had melted into the rock. Once the hands were withdrawn, the impression would ensure vivid proof of contact with the supernatural world.

IT'S EASY TO be seduced by images of a macabre carnival of game drives and mass killings, when in fact much Cro-Magnon hunting involved small numbers of people, sometimes only individuals. The routine of daily life, year in, year out, revolved around the small-scale hunt: the pursuit of red deer at the edges of forests, the quiet trapping of ptarmigan or grouse by women, a couple of arctic hares snared along a trail. The limitations of being on foot, of lightweight, if accurate, weaponry, meant that most hunts required taking not bison or rhinoceroses but deer or ibex, a solitary reindeer or a horse, or unspectacular hours spent tracking arctic foxes during short winter days. Above all, everyone watched animals. If modern-day hunters are any guide, they attributed complex emotions and feelings to them as they watched, sometimes simply out of curiosity.

Hunting larger, more formidable game like an aurochs or a bison was dangerous business and would have required careful preparation. There were always a few such beasts around, familiar denizens of the local valleys to be avoided because of their uncertain tempers and unpredictable behavior. To hunt such a dangerous beast was a major undertaking that would have involved several bands and infinite patience . . .

The massive bull aurochs is a magnificent beast with forward-angled, lyre-shaped horns and a pale stripe down its spine. It grazes alone at the edge of the trees, tail flicking away the summer flies, its black body hard to discern against the dark forest. The hunters and their sons have watched it for days. They track it as it moves down to the stream to drink, snorting fiercely at another bull nearby. They flit from tree to tree in the midday shadows while their prey moves majestically down a narrow forest trail. No one is in a hurry, for they know that their seemingly peaceful quarry can turn aggressive in a moment. It dwarfs even the tallest hunter; its horns can rip open an attacking wolf in a second.

Each night, the hunting party talk about the bull. They dissect its every move, its behavior, its likely reaction to an attack, its daily, and surprisingly predictable, routine. There's a spot along the forest track where it labors through soft brown sand and mud still wet from a spring thaw, a place where trees arch overhead. This is the strategic point for the hunt. Under the direction of a scarred, weather-beaten elder, the hunters collect sharpened antlers and sticks, watch the aurochs out of sight, and then quickly excavate a deep pit in the soft ground with their hands and some reindeer shoulder blades. The boys climb down into the deepening hole and scrabble out the soft clay. Then their fathers set sharp antler stakes in the muddy pit and cover it with branches and then sand. The most experienced hunter climbs onto a branch above the pit with a heavy thrusting spear, while the others hide in the surrounding forest with a carefully folded strong fiber net by their feet. They keep watch in vain that evening. The aurochs stays by the stream, wary of a pride of lions it sees nearby. The hunters wait.

The lions move on, and the aurochs relaxes. Toward midday, the aurochs moves along the familiar track. With a violent crash, it stumbles into the hidden pit. As it bellows and tosses its great horns, the hunter overhead thrusts his spear between the bull's shoulders. His companions quickly throw the net over the raging horns, dancing nimbly away from the flailing head. They hurl spear after spear into its flanks. The tiring beast struggles frantically, but sinks into the churned-up mud. As it slows, the hunters approach cautiously to deliver the death blows . . .

OVER THOUSANDS OF years, the routine of Cro-Magnon life remained unchanged from one generation to the next. Their existence revolved around the cycle of the seasons, familiar environments and landscapes, the customary movements from sheltered winter base camps into summer encampments, and the annual gatherings where neighboring bands came together for a few brief days. Experience counted for everything, the cumulative lore of generations passed from father to son, mother to daughter, knowledgeable hunter to neophyte. The passage of time was cyclical, too, marked by the constant realities of birth and death, by abundance and scarcity, and, above all, by the mass hunts of summer and fall. As long as these hunts continued, survival was ensured. The world of the ancestors and the living continued to pass on to those still unborn.

Mass hunts may have been seasonal events, but they provided about the only way in which Cro-Magnons could harvest large quantities of meat and hides in a short time, essential for survival through long winters. To acquire a rich harvest, they made sophisticated use of the terrain within their hunting territories. The hunters were well aware of the limitations of their short-range weaponry. To overcome them, they became masters at using terrain as natural corrals, at preparing ambushes near fords and swamps, in narrow culs-de-sac and small valleys. It was no coincidence that some painted caves, like Pech-Merle, lay close to deep canyons where horses and other animals could be ambushed. In northern Spain, hunters would prey on ibex and deer at the choke points at the heads of river valleys and their tributaries. Such hunts had an immensely long history throughout Europe, although we lack much evidence of them in the West. As early as thirty-nine thousand years ago, and probably earlier, hunting bands at Kostenki and Borshchevo, in the Don Valley, were driving horses and reindeer into side ravines in regular mass hunts.[10]

Like reindeer, horses were an important source of bone marrow, fat, hide, and meat.[11] Their hooves could serve as small containers or be boiled to extract a form of glue. Late Ice Age horses were very different from domesticated forms. Aggressive, agile, and fierce, they were compact beasts whose closest living relative may be the Przewalski horse of Mongolia. Now extinct except in zoos, the latter was first observed by Colonel Nikolai Przewalski in 1880 and last seen in the wild in 1969. With their large heads, dun coats, and white bellies, Przewalskis closely resemble the horses depicted on late Ice Age cave walls, which often display the characteristic pendulous belly and sometimes even feature the striping found on these beasts (see figure 11.1). As Cro-Magnon hunters well knew, the best way to kill the nimble animals was by using the landscape.

This would not have been a matter of finding a strategically placed cliff and stampeding a herd to their deaths, as Plains Indians did with bison. You can stampede a bison herd by using skillful driving maneuvers that send the leaders into a panic, setting off a headlong flight. When the front runners reach the cliff face, the sheer weight of the herd behind them pushes dozens of individuals over the edge. Wild horses behave very differently. We know from studies of feral herds that when confronted with danger, individuals break away sharply, flee in single file, or just scatter. The only way you can kill a horse herd is by confining the beasts within a box canyon or small valley, which is exactly what Cro-Magnon hunters did at Solutré, near the modern city of Mâcon, at the southern edge of Burgundy in central France.

Each spring and summer, horse herds migrateed between their winter grazing grounds, in the floodplain of the Saône River, and the cooler, insect-free foothills of the Massif Central, where they spent the warmer months and could find plentiful snowmelt for watering. They passed through a valley that served as a natural corridor past a conspicuous landmark, the Roche de Solutré, which forms part of an escarpment sheltered from northwesterly winds. Here the herds paused to graze. Here, too, the hunters lurked in an unchanging routine of slaughter that repeated itself year after year between thirty-two thousand and twelve thousand years ago.

The south slope of the limestone ridge at Solutré is relatively gradual but broken terrain, ideal as a form of natural corral that was a cul-de-sac in the higher ground of the escarpment. After days of watching, the hunters would have decided on a strategy for the kill. Perhaps they herded the beasts with a drive line, gently waving hides to steer them, or lit a row of fires, or even constructed drive lanes marked by stone cairns (now long vanished). It was a question not so much of driving the herd but of pointing it in the right direction, a task that required careful placement of men, women, and children; taking account of the wind direction; and having

the hunters in place in the cul-de-sac. Many generations of cumulative experience went into the Solutré horse hunts, especially into anticipating
the behavior of the trapped horses. The waiting spearmen knew the
spooked herd would scatter in different directions, only to be hemmed in
by the terrain. They could pick them off quickly in the confusion and dispatch any terrified beasts that tried to climb the rugged slopes. Their
strategies worked successfully for over twenty thousand years.

The first archaeologist to work at Solutré was Adrien Arcelin, a local
archivist who found a weathered flint spearhead while walking beneath
the escarpment with his parents in 1866. He became obsessed with what
he believed was a Stone Age kill site. Arcelin dug deep trenches into the
cul-de-sac and under the imposing cliffs of the Roche de Solutré. Influenced by romantic accounts of Plains Indian bison drives, he speculated
that Stone Age hunters had driven their prey over the cliffs high above.
He was soon proved wrong. No bones came from the base of the Roche,
only from the niche-like cul-de-sac on the western side, where stone
projectile points abounded.

Four long excavation campaigns began with Arcelin's investigations
and culminated in a meticulous investigation involving a team of scholars led by Jean Combier and Anta Montet-White, between 1968 and
1998.[12] The Combier and Montet-White excavations focused on bone-
filled layers undisturbed by earlier excavators. The team obtained radiocarbon dates and enough material for a long-term study of what was,
by any standards, an impressive kill site with a long history. We now
know that Neanderthals were the first to kill horses here, about fifty-five
thousand years ago. Then there was a gap of nearly twenty millennia,
represented by over six feet (two meters) of sterile soil, before Aurignacian bands arrived about thirty-four thousand years ago. They were
sporadic visitors until about twenty-nine thousand years ago.

By twenty-eight thousand years ago, Gravettian hunters had inaugu-
rated a pattern of seasonal killing that was to persist for thousands of
years during a period of rapid climatic shifts leading up to the Last Glacial Maximum. Most of the time, the climate was cold, the escarpment
surrounded by the kind of open steppe favored by wild horses and reindeer. Overwhelmingly, the hunters targeted horses. The bands descended

on the kill site between May and November, apparently preferring young stallions, whose bones abound there. Each year, the hunters would kill indiscriminately, slaughtering far more animals than they needed in brief orgies of killing. By twenty-five thousand years ago, Solutré had become a massive kill site, with deep layers of horse bones lying under the surface, even paving the ground. At least thirty thousand horses perished in Cro-Magnon hunts here, many of them unnecessarily. We know this because the packed bone deposits yield entire backbones and back legs, many of them with few signs of butchery with stone knives, as if the hunters simply took what they could immediately eat or dry, then left dozens of carcasses to rot.

We can imagine the Solutré kill site littered with whitened bones when the hunters arrive, dodging the hundreds of beasts grazing near the cliffs. Then days of careful preparation culminate in the frenzy of the hunt. The short grass becomes trampled and bloody, dead horses limp on the ground, others flailing helplessly in their death throes. Now the butchery commences. Men and women swarm over the beasts, gutting them rapidly, cutting out their tongues, and skinning them for their hides. They quickly remove tendons as well. There is meat in abundance, far more than the people could ever eat, but time is short, given the hovering predators. So they cut off strips of flesh to dry and perhaps choice leg bones for their marrow and immediate consumption, then leave the rest to rot. They move away, oblivious to the flies that swarm over the carcasses and to the hyenas that move in, out of spear's range. Within a few days, the accumulation of rotting flesh and bones is all that remains of the kill, the remnants of a promiscuous slaughter that can be smelled miles downwind.

The hunts continued until the increasing cold of the Last Glacial Maximum drove the hunters southward into warmer environments. By 21,500 years ago, the extreme cold had made Solutré and most more-open environments untenable. The hunts ended, as the Cro-Magnon population of the West shrank until about 19,000 years ago, when a slightly warmer and more humid interval allowed Cro-Magnons to return to Solutré. It's a measure of the severity of the climate that this time they hunted reindeer, not horses.

CHAPTER 11

The Magdalenians

LASCAUX, FRANCE, EIGHTEEN THOUSAND years ago. The row of fat lamps at the foot of the wall flicker in the gloom and cast deep shadows among the crevices in the rock. Flaming brands stuck between boulders throw brighter light on the wall that will serve as the artists' canvas. Two men stand on rough ladders lashed together from small branches. They jam saplings into nearby cracks, then lash crosspieces and floor supports to them to form crude scaffolding. The wall bears faint traces of earlier paintings: an aurochs, a horse, and a maze of engraved lines.

One of the artists clambers off the floor and perches on a crossbeam. He brushes the wall with the arm of his parka. A girl passes up two hide pouches and some animal-hair paintbrushes. In one of the pouches, she has carefully mixed water and red iron oxide from outside the cave to form his paint. He inspects the paint suspiciously, then dips the brush and draws the bold outline of an aurochs head on the wall. The massive head is first, then the forward-facing, lyre-shaped horns vivid in his memory from a close encounter with a large bull only a few days before. He calls softly to the girl and passes down paint and brush. Now the artist opens the second pouch, containing powdered black pigment, and pours some into the palm of his hand. With infinite care, he blows the powder onto the rock face, inside the great bull's head . . .

The artists have worked inside the cave since time immemorial, the time of the ancestors. They've returned there again and again, sometimes freshening an existing figure, but more often painting or engraving new ones. Each depiction has a story, a message, and a relationship with the man or woman who painted it. The cave is a quiet sanctuary, sometimes evoking terrifying but usually unfounded fears of slumbering bears. More often it is a place of ceremony. No one has ever lived in this sacred place in the dark, only occasionally visited there, and then only to communicate with the realm of the supernatural in chambers that echo and resonate to chanting voices. On these occasions, pine-resin brands and fat lamps flicker in the gloom, sometimes held close to the animals that live on the walls, causing them to move and jump in the dim light. Here the people connect with the powerful forces of the supernatural, with the animals that surround them on every side. There were many such sacred places in the Cro-Magnon world, but one of the greatest was Lascaux.[1]

FOUR BOYS FOUND the cave near Montignac, in France's Dordogne, in 1940, when their dog vanished down a burrow while hunting rabbits. They heard him barking underground, fetched a ladder, and found themselves confronted by giant bulls. Lascaux had been sealed since the late Ice Age, so what the Abbé Henri Breuil soon called "the Sistine Chapel of Prehistory" was intact.

Lascaux overwhelms the senses (see color plates 6 and 7). Seventeen horses, eleven aurochs, six stags, a bear, and a mythic beast encircle the walls of the first chamber. They call it the Hall of the Bulls, dominated as it is by two giant aurochs, which face one another at a larger than life scale. The left-hand bull is unfinished, his head covered with brown spots. The right one is a vast beast, nearly 15 feet (4.5 meters) long. He seems to be snorting and is, perhaps, about to paw the ground. A row of horses gallop across the wall immediately below and behind the left-hand aurochs, followed by a strange beast that some call a unicorn. Perhaps it's also an aurochs. There are stags, too, with elaborate racks of antlers. All the paintings were executed in polychrome and were painted at different times—the horses first, followed by the aurochs, then the stags. Rock-art expert Norbert Aujoulat has studied the subtle indicators of weight and coat colors. He believes the artists painted the horses as they would appear at winter's end and in early spring, the aurochs in their summer coats, and the stags in autumn finery, the very seasons at which each mated.

The Hall of the Bulls opens into the smaller Axial Gallery, crowded with at least fifty-eight animals and numerous symbols. A large black stag with branching antlers appears to bellow at the entrance to the gallery. A great black bull also faces the entry. The artist executed much of the bull by blowing pigment on the wall, then delineating many details of the head and legs with a brush. Three prancing horses with yellow or brown bodies give way to a group of five. In another panel, two brick-red-and-black male bison with bristling manes stand back-to-back, slightly overlapping and painted at the same time as a deliberate composition. Six hundred paintings of animals and signs and fifteen hundred engravings, mainly of horses, adorn the walls of the cave, many of them composed atop one another (figure 11.1).

Lascaux has a lower network of galleries, which once had its own entrance. At the end of this network lies a small chamber—now only accessible from the main part of the cave—where many generations of artists worked. One of them painted a scene, one of the few in Cro-Magnon art. A charging bison with lowered head, threatening horns, and raised tail confronts a falling human figure with outstretched arms, erect penis, and bird head. A barbed weapon has pierced the bison's

Figure 11.1 *A horse from Lascaux. Robert Harding Picture Library/Alamy.*

hindquarters, which have been disemboweled. A bird on a stick, perhaps a spear thrower, stands below the falling bird-man. Scholars argue endlessly over the significance of this scene. Does it represent an actual hunting incident, or is this a symbolic depiction of the passage between life and death, where the bird symbolizes the flight of the soul from the dead? Rock-art expert Jean Clottes points out that the location of the scene is a place that is very prone to high concentrations of carbon dioxide, which can cause both discomfort and hallucinations.

No one knows exactly when Lascaux was painted, but it's clear that the cave was in use for many centuries. Unfortunately, the paintings were made with mineral pigments, which cannot be radiocarbon dated, so we have to rely on dates taken from artifacts found in the cave, which appear to run between nineteen thousand and seventeen thousand years ago. We must assume that the artists worked there once, on many occasions, or over thousands of years. The chronology is too imprecise.

THE LAST GLACIAL Maximum held most of Europe north of the Mediterranean and the Pyrenees in such a frigid grip that many Cro-Magnon groups retreated into sheltered valleys or moved southward into warmer locations. Then, about nineteen thousand years ago, conditions warmed significantly during a brief interval of one thousand years or so often called the Lascaux Interstadial. The Solutré kill site is a barometer of what happened. Apparently, both animals and humans abandoned the area after twenty-four thousand years ago, then returned during the brief interstadial, which is marked by soil formation and erosion over a wide area of southwestern France and northern Spain.

It was as if nothing had happened. The seasonal killing at Solutré resumed, but now the prey was reindeer rather than horses.[2] This time, too, the hunters used not only bone-pointed spears but also weapons bearing what French archaeologists rather elegantly call *feuilles de laurier*, "laurel leaves" (figure 11.2). These beautifully made stone projectile points do indeed look like idealized laurel leaves and stand out as exotic in otherwise unchanging tool kits of bone artifacts, burins, and scrapers. Those skilled enough to fabricate them had mastered a new stoneworking technology, which involved using an antler billet to squeeze off shallow flakes by applying sharp pressure along the edges of a blade. This technique—pressure flaking—produced thin, beautifully shaped yet functional spear points that were both lethal and lovely to look upon. Sometimes, the stoneworkers made what one might call rudimentary versions of the points using pressure flaking on but one side of the tool. On occasion, too, they made spearheads with a shoulder that served as the mount for the shaft. But the ultimate was the classic laurel leaf, flaked on both sides, beautifully regular and thin. *Feuilles de laurier* were never common, and indeed, some researchers wonder if they were, in fact, ceremonial tools and never used in the field. This seems unlikely, for they would have made tough, effective weapons for killing prey like reindeer.[3]

The Solutreans—to use the archaeological term—who made these tools are shadowy folk, whose only claim to fame is their distinctive spearheads. We know from the bones in their occupation levels that they were hunting cold-loving animals between about 21,500 and 17,000 years ago, at the height of the Last Glacial Maximum. Small numbers of them still based themselves in the sheltered valleys of southwestern France and in the foothills of the now-glaciated Pyrenees, but many bands flourished in the diverse, warmer environments south of the mountains. There were far fewer of them than there had been Gravettians. The *feuilles de laurier* were the only exotic artifacts made by these people, whose lives and hunting practices were otherwise identical to those of other Cro-Magnons of the day. Solutrean hunters always used spears armed with backed blades and bone heads, too, and they became more common as time went on, until the *feuille de laurier* finally vanished about 17,000 years ago. Perhaps the Solutrean is simply a reflection of a brief adjustment to slightly different hunting conditions, where stone-tipped thrusting spears assumed greater importance for a while. We will never know.

If the Lascaux chronology is to be believed—and remember that the radiocarbon dates come from artifacts in the cave, not actual paintings—then Solutreans were the artists who painted there, which gives us a quick snapshot of the diverse bestiary that surrounded them.

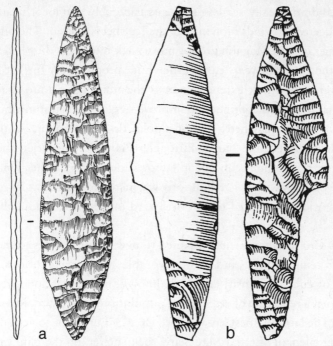

a   b

Figure 11.2 *Solutrean hunting weapons. (a) Laurel leaf point (French: feuille de laurier) (b) shouldered point. After P. Smith, 1966.* Le Solutréen en France. *(Bordeaux: Imprimerie Delmas).*

Aurochs, bison, horses, and stags dominate the walls, though not reindeer, which were the staple of the time. It seems, however, that much of their spiritual life surrounded larger animals like the wild bull, which was an aggressive, frightening prey, as was the bison, commemorated in the death scene there. We do not know whether the Solutreans' ceremonial life and beliefs were similar to those of their predecessors, but it seems certain that there were close spiritual and supernatural links between humans and their prey, just as there had been in earlier times.

THE FIVE OR six millennia after the Lascaux Interstadial witnessed a gradual climatic amelioration that proceeded in fits and starts, with quite rapid warming at times. Ice sheets covering the Alps and Scandinavia retreated dramatically, sea levels rose considerably, and loess accumulation triggered by winds blowing off the ice sheets slowed. The ongoing warming saw Cro-Magnon groups move back into areas they had abandoned thousands of years earlier, then hunt progressively further north into river valleys that extended across the tundra and onto the open plains as conditions continued to improve. A classic environmental phenomenon was in play here, that of a polar desert breathing. As the climate deteriorated, the North European Plain became drier and less habitable. Both animals and people moved out to its southern margins and beyond. However, as conditions warmed and ice sheets retreated, the less arid land sucked them northward into hitherto uninhabitable lands.

Now Cro-Magnon society adapted to new circumstances, somewhat warmer temperatures, and more-favorable hunting conditions. This period of cultural transition is little known, for occupation layers are transitory, a reflection of still-sparse populations and very cold temperatures. The last, and best known, late Ice Age European society, named the Magdalenian after La Madeleine rock shelter, on the banks of the Vézère River near Les Eyzies, emerged during this period, almost certainly from Solutrean roots. This was probably a time of tentative advances, and sometimes retreats, across more-open terrain, as always following the herds of horses, reindeer, and other game that had been the foundations of human life for thousands of years (see color plate 14).

The earliest Magdalenian sites may be those in the Cantabrian region of northern Spain, which date to before seventeen thousand years ago. Within a relatively short time, some groups had settled in a small area centered on the Périgord region of southwestern France about 124 miles (200 kilometers) across. We know little about these sites' inhabitants, except what can be gleaned from their ubiquitous and not particularly distinctive stone artifacts, especially their many burins. The chisels are hardly surprising, for each band made heavy use of short, squat antler and bone projectile points, most with simple beveled bases (figure 11.3). Such weapons were ideal for hunting in relatively open terrain, especially when used with spear throwers. Needles are common finds. Temperatures were low, winters long and harsh, tailored clothing essential.

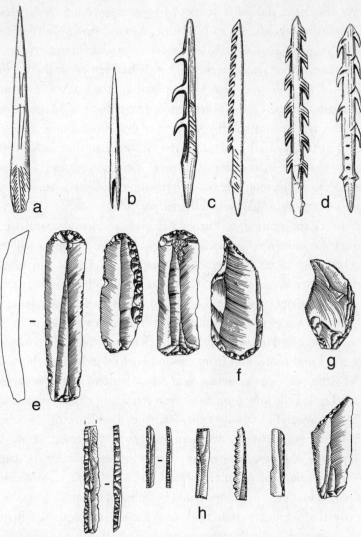

Figure 11.3 *Magdalenian tools and weapons. (a) Beveled bone point. (b) Forked base point. (c) Single and double barbed harpoons. (d) End scrapers. (e) "Parrot-beaked" burins. (f) Small blade tools, probably mounted in handles. (a, b, d, e, f): After Jean-Luc Piel-Desruisseaux, 2007.* Outils Préhistoriques *(Paris: Dunod). Reproduced with permission. (c) After M. Julien. 1982. "Les harpoons magdaleniéniens," XVIIe supplement,* Gallia Préhistoire.

Over about five thousand years of irregular warming, Magdalenian populations rose gradually, to the point where some favored localities, like the Vézère Valley, may have become quite crowded by sparse late Ice Age standards. By 14,500 years ago, bands had settled in the northern foothills of the Pyrenees and were spreading gradually over a much wider area as climatic conditions improved rapidly. The Magdalenians' technology remained much the same, with ever more use of antler and bone for artifacts of all kinds. Like the Gravettians, they were efficient hunters, expert with the spear thrower. Increasingly, they fabricated smaller stone tools; many served as barbs on ever-lighter weapons used in open country and against smaller game.

About 14,500 years ago, Europe's climate suddenly warmed in an amelioration known to paleoclimatologists as the Bölling oscillation, which lasted for about 1,500 years (see chapter 12). The density of Magdalenian settlements increased rapidly during Bölling times, as Cro-Magnons came out of shelter. Throughout the Last Glacial Maximum, the sparse human populations north of the Pyrenees had spent most of the year in sheltered locales like deep river valleys, where there was some forest cover and protection from chilling north winds. The short summer was the only season of the year when hunters could work freely outside. During the long months of below-freezing temperatures, activities in and around their base camps depended on the cold. We should not forget that modern-day northern hunters are reluctant to take long hunting trips when the temperatures drop below minus thirty degrees Fahrenheit (minus thirty-four degrees Celsius). For the Cro-Magnons, traveling on foot even in groups could be lethal in the cold, and even the most trivial of accidents could be deadly. The survival equation changed rapidly with the Bölling's rising temperatures. Winters, while still intensely cold, were somewhat shorter, summers longer. With fewer climatic extremes and more-favorable traveling conditions, the Magdalenians expanded rapidly into hitherto inhospitable landscapes. Many groups hunted horses, reindeer, and other tundra species as they settled on the harsh open plains to the north, from the Paris Basin to Belgium and Germany; moved into the Rhine and Meuse valleys, as well as Britain;

and settled at higher altitudes than the Dordogne in France's Massif Central, the Alpine foothills, and the Pyrenees, at altitudes of sixteen hundred feet (five hundred meters) or more, unthinkable only a few centuries before.

As temperatures rose and summers became slightly longer and warmer, Magdalenian groups expanded into long-empty landscapes. By fourteen thousand years ago, Magdalenians were killing reindeer at Solutré, apparently using the same strategies as their predecessors. Other groups ventured far northward, some of them to the Ile-de-France, in the Seine Valley, in what is now the heart of Paris, far from the shelter of southern river valleys. Hunting territories were large, centered on locations where horses or reindeer could be intercepted or where larger animals like bison could be mired. The people would return to the same campsites for many generations, living in tents pitched in places where huge piles of whitened reindeer and horse bones formed veritable monuments to long-forgotten hunts. Here on the tundra, bone lingered and seemed to never rot away. Forests of skulls and antler racks surrounded the tents, impervious to the depredations of hyenas and lions. The campsites themselves must have sat on pavements of broken bones torn to shreds by humans and other predators. However, today, all that survives are stone scatters, boulders that served as the anchors for tent walls, and charcoal from large hearths.

The tents would have had to be durable, waterproof, and easily made. Judging from historic arctic summer dwellings, some were dome shaped, others probably cone shaped, perhaps about 15 feet (4.5 meters) across and 10 feet (3 meters) high.[5] Rough wooden poles would have provided a frame, covered with reindeer hides stitched together while still green and pliable. The covering would have looked like a patchwork of crude stitching, the seams pulling apart and leaving gaps as the hides dried (see color plate 15). A circle of boulders would have held the bottom down, for wooden pegs would have broken before penetrating the permafrost. Dried hides would have provided a door. These were very much summer dwellings, although a second layer might have provided winter protection. We can surmise that semisubterranean houses with mammoth-bone, timber, and sod roofs, like those used by eastern Gravettians, came into use during the cold months, but none have yet been discovered in the West.

While the men hunted, the women remained in camp most of the time, not only preparing food but also devoting enormous amounts of time to making clothing and preparing hides and furs. They would have treated the skins of all land animals by wetting, sometimes with urine, then scraping them, which is one reason why stone scrapers are so common in Cro-Magnon archaeological sites. Hours of hard scraping removed fat from the skin and made it supple. Then the skin was thoroughly wetted again and rolled up, with the skin side outward, for a day or more. Next the tanner laid it out, felt it carefully, and scraped any hard spots away before rubbing the scraped (hairless) side with a mixture of sand and water. Finally, she hung it out to dry a short distance from a fire, rubbing it again when it was warm. For a hairless hide, the tanner would have soaked it in urine or water for about four days to loosen the hair. A pervasive smell of urine no doubt surrounded the camp, accompanied by the endless scrape-scrape sound of stone against hide.[6]

Cro-Magnon women must have tanned the skins of a wide range of animals, large and small, including bear hides, hare pelts (scraped, coated with a curing mixture, then dried), and bird skins, which would have been scraped, probably with a bone scraper, then treated with fish oil before being softened by being rubbed to and fro. Skinning and tanning must have gone on on year-round, but the most intense tailoring would have taken place in the weeks before the fall reindeer hunts, before temperatures dropped precipitously. Magdalenian women's work was never done, for they played as important a role in daily life and survival as the men.

THE BÖLLING WAS a world of dramatic environmental contrasts: of unusually warm summers and exceptionally mild winters that then gave way to years of extreme cold.[7] Life was a seesaw of changing climate in which ancient hunting routines continued as before but environmental shifts meant changes in prey, sometimes from reindeer to red deer, or to wild boar. Increasingly, the Magdalenians hunted many smaller animals, especially birds and fish. As the Atlantic rose and flooded the continental shelf, the gradients of rivers like the Dordogne and the Garonne shallowed. Seals appeared in local waters, while huge salmon runs crowded strategic rapids in spring and fall. Engravings of seals and salmon appear on Magdalenian artifacts of the day. In one instance, the latter cavort among the legs of swimming reindeer . . .

The chill river water of spring runs brown and fast, tumbling among the black rocks and rapids where you can cross dry-shod later in the summer. Soon large salmon will crowd the shallows and pools, leaping high as they run upstream to spawn.[8] They arrive without warning each year with the spring thaw, sometimes so many of them that they jostle one another for space. Despite the lingering cold and occasional snow showers, several bands have set up camps nearby in anticipation of the fish. They return to the same locations year after year, erect their hide tents, repair the drying racks still standing from the year before. Children stagger under heavy loads of firewood for the smokehouses; their mothers strike off piles of flint blades to use as gutting knives. Leather bags of antler harpoons lie by piles of fishing spears, ready for the day when the salmon will appear.

The run begins without warning. As the first fish arrive, the fishers climb onto convenient boulders at water's edge, close to stout wooden fences in the river that were erected the previous autumn when the river was low. The salmon crowd toward the barrier's gaps. Effortlessly, the spearmen clasp their weapons close to the point of balance, between the thumb and first and second fingers. With deft, short arm movements, they thrust at the fish, the spears never leaving their hands. Some fishers use spears with two barbed heads that face one another. They impale their prey and quickly flip them ashore. Others work with simple barbed spears, the head attached to the shaft with a line so that the fish can be hauled ashore if need be. The men work at lightning speed, landing salmon after salmon with an economy of movement born of long experience. An antler point snaps inside a large fish as it lands on the bank. In a few moments, its owner removes the fractured head of the spear and binds a new one in its place. Some of the fishers work more slowly, watching their sons as they practice their casts, offering encouragement, sometimes helping with a heavier fish. Still more hunters hover near the barrier and deep pools with circular nets set on poles, catching jumping fish in midair or scooping them up as they swim near the surface.

The work has hardly begun. Within seconds, a man or woman clubs each landed salmon and frees the spear. Away from the river, they rapidly gut the catch and butterfly the carcasses, then suspend them on sticks or across racks around or above slow-burning fires, or in places where the fish will dry in the spring sun and wind. The bands work day and night, flaring pine brands casting flickering shadows over the teeming river. They know they have only a few days to exploit the run, which will end as suddenly as it began. By the time it is over, the harvest will have yielded many hundreds of salmon, most of them to be dried for use during the cold months . . .

The salmon fisheries relied on technologies of drying, smoking, and trapping that had been in use for thousands of years. Exploiting the runs to the full required new, more specialized weaponry such as the barbed antler spear point and harpoon, crafted with one row of barbs, then later with two.[9] Such artifacts are most common at sites near salmon rivers, at places like the Laugerie Haute rock shelter, close to the Vézère. Here generations of visitors combined fishing with harvesting migrating reindeer herds at a nearby ford. Complete reindeer skeletons lie between the cliff overhang and the river. On a nearby rise, the hunters dug pits and filled them with bones and stone artifacts. Perhaps these were blinds where they waited for their prey or or places where they stored fresh carcasses during the spring or fall harvests.

Magdalenians continued to trap fur animals and birds with the same simple snares and traps employed by their ancestors. They may also have developed new weaponry: the bow and arrow. The bow has a major advantage in that you can fire arrows upward, downward, at every angle imaginable, in ways that would be impossible with a spear. Unfortunately, no one has found a wooden bow in any late Ice Age Cro-Magnon site, so we can only guess at when such weapons came into use. You certainly cannot identify the use of arrows from the size of stone barbs, for spear throwers also depended on lightweight spears for their effectiveness. Most likely, bows developed as a forest weapon, at first perhaps little more than a bent sapling with a fiber string designed to shoot solitary animals like deer. As arrows became lighter and ranges improved, expert hunters likely started shooting birds on the wing or waterfowl on ponds, where they could wait downwind for the carcass to float ashore.

Many Cro-Magnon experts believe that some of the earliest bows were used about 19,500 years ago at the Parpalló cave, in the Valencia region of Spain, a site where numerous fine projectile points have come to light. From these warmer, more forested locales, the bow and arrow may have passed northward as temperatures warmed. As we'll see in chapter 12, we know for sure that the bow and arrow was in widespread use immediately after the Ice Age. However, it seems almost certain that the Magdalenians of earlier millennia were at least experimenting with the new weaponry in a changing world where smaller game and birds were slowly assuming greater importance.

AT FIRST, MAGDALENIAN populations would still have been small, many of them concentrated in sheltered river valleys and near strategic fords where game could be found. However, despite their small numbers, contacts between neighboring bands were probably more frequent than in earlier times because there were more people scattered across the landscape. Inevitably, tensions must have arisen. There may have been quarrels over hunting territories, over control of stone outcrops, or between different kin groups. Almost inevitably, too, there may have been sporadic violence, but we have only one find that documents it.

The Cap Blanc rock shelter, near Les Eyzies, contains a frieze of animal sculptures carved by Magdalenian artists about fifteen thousand years ago (figure 11.4).[10] Horses cavort across the wall, the largest more than 6.5 feet (2 meters) long. There are bison and deer heads as well. The artists painted the sculptures in bright colors, so that they were visible from some distance. The grave of a young woman lay in front of the central horse, an ivory point in her abdominal area, possibly the cause of her death. Perhaps this woman was one of the artists, buried in front

Figure 11.4 *A wild horse from Cap Blanc, France. Courtesy: Walking Dordogne.*

of her work. She is the only known victim of what must have been sporadic violence, triggered by competition over hunting grounds, kin rivalries, or competition for marriage partners—to mention only a few potential causes.

The rhythm of the hunt, of survival, continued without much change during the warming time, except for the development of even-more-sophisticated, often multipart hunting weapons and an increasing focus on smaller mammals, birds, and fish. The landscape and environment were changing, coniferous forests more widespread. Magdalenian hunters had considerable leisure, which they spent watching game, making and repairing weapons, and acquiring intelligence through conversations with others. We know that they had more spare time than their predecessors, for they lavished their spears, spear throwers, and other antler and bone tools with engravings that probably had no particular symbolic importance. The surviving art objects that have come down to us are a fascinating mixture of doodles, patterns, and, occasionally, works of genius (see color plate 10).

The Aurignacians, Gravettians, and Solutreans had decorated antler and bone objects with geometric designs and crafted female figurines and other sculptures. The Magdalenians took personal decoration to a new level. I say "personal" because the objects they lavished with talent were items that were in everyday use.[11] The enigmatic *bâton de commandement*, an antler beam with one or more perforations, is one such artifact, probably used to straighten soft antler and bone for projectile points, just as a similar artifact was so used by the historic Inuit (figure 11.5). (The strange term *bâton de commandement* implies there were powerful authority figures, even kings, among the Cro-Magnons, not a reasonable assumption.) The owners of the *batons* sometimes carved them with animal heads, engraved a baying stag with his magnificent autumn antlers, or wrapped reindeer antlers and abstract designs around the shaft and hole. They weighted spear throwers with exquisite deer and mountain goats perched on the hooked end (see color plate 10). The animals are defecating, and the excrement serves as the hook. Sometimes, too, a bird feeds off it. Another, alas broken, spear thrower is weighted

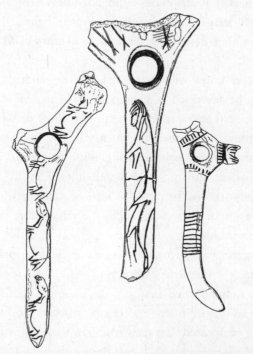

Figure 11.5 *Magdalenian bâtons de commandement. After Jean-Luc Piel-Desruisseaux, 2007.* Outils Préhistoriques. *(Paris: Dunod). Reproduced with permission.*

with a bison turning its head backward and licking its flank. Even the tear duct of the eye and the mane are delicately engraved with assured burin strokes. Some of these masterpieces may have been for ceremonial use only.

Lengths of bone and antler bear circular and geometric motifs, which may, perhaps, be copies of painted designs on people's skin or adorning hides or wood. The Magdalenians were consummate doodlers on antler and bone, ivory and stone: of animals of all kinds, even of occasional human figures, and sometimes of palimpsests of designs superimposed on one another that take hours to decipher. Among these many doodles, the work of just about anyone, are some true masterpieces. A line of felines stalks along a bone. A cutaway engraving of a horse on both sides of a flat bone fragment is adorned with what almost looks like a halter, but probably is not. There are bison and horses, reindeer and stags, even birds, in a seemingly endless parade of animals, as well as abstract designs. And sketches of human figures abound both on small objects and on cave walls, where the full glory of Magdalenian art can be enjoyed.

We will never know why the Magdalenians suddenly embarked on what can only be described as an orgy of artistic creation at every level, from the mere scribble to true masterpieces, some of which must have had ritual significance. I suspect that it was a reflection of a warming world where populations were growing, hunting territories were becoming more crowded, and group and individual identity was assuming ever-greater importance. It may also have been a function of extended occupation of major sites. Sometimes the inhabitants laid stone platforms inside their hide tents at open campgrounds, as they did at Plateau Parrain, in France. Except for the Sungir burials described in chapter 8, we have no direct sense of the elaboration of Cro-Magnon society, of the bright colors that adorned people's faces, of the feathers and headdresses that identified hunters and elders, shamans and fellow kin. We have strings of perforated bear teeth and shells, some of them from as far away as the Mediterranean coast, which must have had considerable exotic value. This was a society, or, more likely, a series of societies, that maintained sporadic contacts with groups living significant distances away, some of them in quite different coastal, forested, and highland environments. Such contacts made sense in a world where game covered enormous distances and winter food shortages were commonplace.

For all their artistic genius, we know nothing of the full pageantry of the Magdalenians' life: the colors, the smells, the chanting, flutes, drums, and dancing that must have been an important part of what was, undoubtedly, a rich and essential ceremonial life. In winter, people must have stayed closer to home, crowded into small dwellings, where the long nights passed in storytelling, in songs and chants. These were the intimate months, when life revolved around family and kin, when tensions flared in confined spaces and qualities of tolerance and humor came to the fore. Most likely, important ceremonies unfolded during the warm months, when bands came together for a few weeks.

All these activities were aboveground, familiar rituals and storytelling that took place in camps and rock shelters. But there was also an underground dimension, often far from daylight, where temperatures were always constant and the supernatural realm lurked behind cave walls. It is underground that we have a glimpse of Magdalenian genius, and of the complex spiritual beliefs that were a buffer between the living and the powerful forces of the supernatural.

WE KNOW THAT the Cro-Magnons ventured far underground, into dark, remote, and probably terrifying places. Every time they went into a cave in winter, they faced the threat of hibernating bears, for many caverns were also dens. For generations, researchers assumed that the only visitors to the depths of caves were men and boys, who went there for rituals such as initiation ceremonies. We were wrong. Close study of footprints deep in the caves tells us that people of both sexes and of all ages went into remote chambers belowground. Even the handprint of a baby lies deep below the surface in the Bédeilhac cave, in southern France.

At the Tuc d'Audoubert cave in the Pyrenees, visitors plucked teeth from the skeletons of cave bears that had once hibernated there, presumably for amulets hung around the neck or on clothes. Deep in the cave, nearly three quarters of a mile (one kilometer) from the surface, two three-dimensional clay models of bison lie in a small chamber, deliberately set against a rock (see color plate 11).[12] The rear beast, a male, appears to be about to mount the female with raised tail in front of him. Each was molded from lumps of clay dug up from the floor of a neighboring chamber. The footprints of a young child survive close by. Some deep caverns were difficult to access. At the Fontanet cave, in Ariège, in southern France, the Magdalenians crossed a chasm into the deepest part of the cavern by clinging to wall fissures. Numerous foot- and handprints survive on the clay floor, among them that of a right foot whose blurred imprint is such that the owner may have been wearing some form of footwear, probably made from hide.

Magdalenian artists were sculptors and painters, expert engravers and doodlers. Cave paintings from this general time period occur in the Urals, far to the east, and as far north as central England, but the finest are in southwestern France and northern Spain, in such well-known locations as the deep cave at Rouffignac; Font-de-Gaume and Les Combarelles, near Les Eyzies; and a series of painted caves in the Pyrenees and Cantabrian Spain. And the ultimate masterpiece is Altamira.

The Altamira cave, near Santander, near the Biscay coast, is 961 feet (263 meters) long, a cavern of chambers and passages, ending in a narrow defile known as the Horse's Tail.[13] A local landowner, Don Marcelino Sanz de Sautuola, noticed some black marks at the back of the cave in 1876, but thought nothing of them until his eight-year-old daughter Maria, bored with his excavations, wandered with a candle into a side chamber. "*Toros! Toros!*" she cried, in one classic, and seemingly authentic, tale of archaeological discovery. Father and daughter gazed in amazement at the colorful bison on the ceiling of the low chamber.

Sautuola noticed close similarities between the art and pictures of animals he had seen on antler and bone fragments from French rock shelters at an exhibition in Paris. He claimed that the Altamira bison were the work of Stone Age artists but was ridiculed by scholars for his pains. The unfortunate landowner was vindicated after his death by paintings and engravings discovered at La Mouthe and Les Combarelles caves in 1895 and 1901.

The bison chamber at Altamira has a low ceiling, which may be why the artist chose to paint there, taking advantage of the natural bulges in the rock face (see color plate, 12 and 13). The paintings cover the 19-foot (5.8-meter) ceiling, a series of superimpositions that were painted in five phases, beginning with engravings, followed by various colored paintings, then some multiple-line figures, and finally the polychrome bison. The line figures on the ceiling are identical to some found as portable objects in the cave deposits, which have been radiocarbon dated to 14,480 years ago. Charcoal from some of the polychrome bison dates to between 14,820 and 13,130 years ago, at which point the cave was blocked off.

One artist was probably responsible for most of the polychromes, even if the radiocarbon dates hint that others later retouched them. There are twenty-seven bison, four red deer does, one red deer buck, and two horses. The unknown artist painted the animals on natural protuberances from the ceiling to give them bulk and a sense of roundness. He or she was a close observer. Judging from studies of modern buffalo, several of the bison are apparently rolling in urine-impregnated dust with their legs flexed, a common way of marking territory. The bulges in the ceiling give them a three-dimensional effect. Two stand at rest; others bellow with their heads raised. You're entranced by the polychrome paintings and fail to discern the more than seventy other figures that depict a very different suite of animals—including twenty-three deer, five of them males, five ibex, two horses, two aurochs, and eight anthropomorphic figures with birdlike masks. They are hard to see, as if the intent was entirely different from that for the colorful bison. If you look closely at the paintings, especially the directions in which the lines were drawn, you can tell that the artist seems to have intended that the viewer see them only from the entrance of the gallery, where angled light would have made the bison stand out from the ceiling.

Though Altamira's original entrance was blocked some thirteen thousand years ago, you can get a sense of what the cave was like in its heyday by visiting the magnificent and extremely precise replica that lies a few hundred meters away from the original. As rock-art expert Paul Bahn points out, the visitor can now see the decorated chamber in relation to the original entrance and to daylight.[14] We can also inspect the

entire ceiling, but the artist and the original Magdalenian visitors would
have had to crouch between the floor and the low ceiling. They would
only have seen parts of the ceiling at one time.

THE GENIUS WHO painted Altamira's bison lived at a time when the arctic shackles of the last Ice Age glaciation were looser than they had been
for thousands of years. Summers were longer and warmer, water meadows and coniferous forests covered much of the Vézère Valley, and the
southern margins of the tundra were shrinking northward. With the
warming, the great bestiary that had supported the Cro-Magnons for
thousands of years changed imperceptibly from one generation to the
next. When reindeer herds moved northward, herds of red deer replaced
them. Aurochs and bison became more common prey. Soon, as warming
accelerated still further, the ancient rhythms of Cro-Magnon life would
change dramatically.

CHAPTER 12

The Challenge of Warming

NIAUX, ARIÈGE, FRANCE, thirteen thousand years ago. The artist sits on a boulder, his hands still bloody from the hunt. A murmur of excited voices assails his ears as he contemplates the magnificent bison lying below him, bleeding from the deep, fatal wounds in its flank. The butchers wait impatiently, watching the lengthening shadows, wanting to skin and strip the carcass before sunset. He draws the curved line of the dead beast's back, sketches in the spear points projecting from the carcass. He holds the charcoal up to his eye, then quickly outlines the massive head. A few deft charcoal strokes delineate the eye, the legs in the characteristic posture of death. Next the curving horns, the mane and beard, then the nostrils and tail—the elderly painter turns the shoulder blade back and forth in his hands. He examines his sketch from different angles, slightly modifies the forelimbs, thinks of the blank space on the wall of the nearby cave. As the artist gets up and stretches in the warm sunshine, the butchers move in . . .

The Niaux cave lies in the foothills of the Pyrenees amid a cluster of Magdalenian settlements. The once-small entrance is at the foot of a steep cliff, part of the 3,900-foot (1,189-meter) Cap de la Lesse massif, perhaps once an element of a symbolic landscape.[1] The labyrinthine chambers and galleries of the cave extend nearly 1.2 miles (2 kilometers) underground. Cro-Magnon visitors explored every underground defile, but they never lived inside—only a few tools and the remains of a hearth lie there. Some of them dwelled in La Vache cave, on the other side of the valley; there Magdalenian groups camped between fifteen thousand and twelve thousand years ago, hunting ibex and other prey. There is an interesting polarity here between the encampment on one side of the valley and the painted cave on the other.

Some sixteen hundred feet (five hundred meters) inside Niaux, four galleries converge at a crossroads, then lead to the main painted area, the Salon Noir. Sixty-six feet (20 meters) across, the Salon is remarkable for its superb acoustics, which magnify sounds and voices, an ideal setting for dance and song and perhaps the reason why there are so many paintings here. Six unevenly distributed panels display bison, horses, and ibex executed in black with a panache of startling realism (see color plate 16 and figure 12.1). One panel depicts eight bison, two ibex, and the back of a horse shown as a single line, arranged in two columns. One of the bison, impaled with two spears, faces a smaller bison head, perhaps her calf. The position of the larger bison's legs suggests that she is lying on the ground, dead. Another bison, this time a male, at the bottom of the left column and painted with exquisite detail, seems to be standing absolutely still, listening for something. Niaux's artists painted images large and small with a vivid recollection of animals they had encountered outside the cave: including fine details of eyes and hair, horses' winter coats, even individual faces. Elsewhere in the cave, there are human footprints preserved on the floor and engravings in clay of, among other animals, a salmon and a bison, but most of the paintings (except for some enigmatic, indecipherable signs) are in the Salon Noir.

Niaux was a sacred place, used for many generations. Many artists

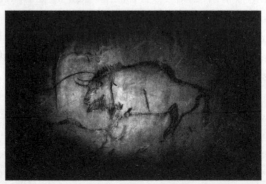

Figure 12.1 *Niaux bison. Sisse Brimberg/National Geographic Stock.*

worked here, even changing their pigment recipes in mid-painting.[2] This is one of the few painted caves with radiocarbon dates from individual paintings. One bison dates to about 13,850 years ago, another to 12,890 years before the present, about the same time as La Vache occupation across the valley. But Niaux's artists worked in a changing world. They were among the last practitioners of an artistic tradition that had endured for at least eight hundred generations.

BY THE TIME of the Niaux bison hunt, Europe was no longer a polar desert. By sixteen thousand years ago, the northern ice sheets were in full retreat across Britain and Scandinavia. The constant winds carried seeds northward, as did birds and other migrating animals. By now, cold-tolerant herbs and shrubs had colonized central Europe's rolling hills. As temperatures continued to rise, woodland spread northward across what had once been permafrosted landscape. Not that the warming was constant. Europe's climate swung like a yo-yo in irregular cycles of colder and warmer conditions, triggered by Dansgaard-Oeschger events about every fifteen hundred years, which brought about brief warming, then prolonged colder times.

Incredible though it may seem, beetles are a reliable barometer of Ice Age temperatures. *Boreaphilus henningianus* was commonplace in Britain during the Last Glacial Maximum and only flourishes in Finland and northern Norway today. During the Bölling oscillation, *Boreaphilus* vanished completely—replaced by beetle species that were much the same as those in Britain today—with summer temperatures averaging as high as 62.6 degrees Fahrenheit (17 degrees Celsius).[3]

By 13,500 years ago, extensive birch, pine, and poplar forest covered Britain, northern Germany, and much of southern Scandinavia in another warm spike, known in some areas as the Alleröd oscillation (figure 12.2).

ALL THESE CLIMATIC shifts were rapid by geological standards, but effectively imperceptible for the Cro-Magnons. The hunters were aware only of subtle changes within a span of a generation: unusually early

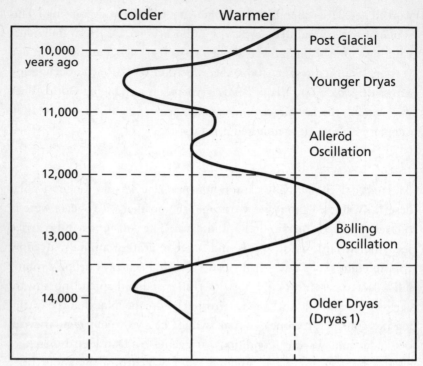

Figure 12.2 *Late Glacial climate change.*

springs, earlier nesting seasons, less snowfall and lusher grazing for horses. The timeless routines of hunting and gathering continued their endless cycles through the seasons. Their lives measured in the brief generations of the day, people still lived the way their grandfathers had and knew that their descendants would have the same kind of existence. Their world revolved around small hunting territories, local river valleys, and the gatherings when people came together and passed intelligence from band to band. As warming continued, much of the talk must have been about changing prey. From the hunter's point of view, the echoes of rising global temperatures brought gradual but profound changes in ways of obtaining food.[4]

As temperatures warmed and woodland spread outward from the sheltered valleys where it had survived even the coldest winters, reindeer migrations shifted as most of the great herds moved northward with the shrinking tundra. The Cro-Magnons had always hunted some red deer, which had thrived in the south throughout the last glaciation. Now red and roe deer, as well as wild boar, became the most common prey. For thousands of years, the hunters had harvested migrating reindeer herds in spring and fall. Such hunts required cooperation, as did the pursuit of large animals like aurochs and bison. As woodland spread and open country shrank, hunting in wooded landscapes changed from an often communal pursuit into a more solitary enterprise. Hunters stalked boar and deer in the forest, using traps and nets as before, but also the spear and, perhaps most important of all, the bow and arrow. They also preyed on a much wider range of smaller animals, all of them familiar from earlier times, but now of greater importance as much of the Ice Age bestiary became extinct or moved northward with the tundra. Rabbits and migrating waterfowl were valued foods in diets where people ate far more plant food than the mere cupful or so they had consumed each year in much colder times.

Such was the flexibility of Cro-Magnon life that the changeover to warmer landscapes was effortless, even if it involved major adjustments in ancient routines. Altamira and Niaux were among the last great painted shrines used by Cro-Magnon hunters for many generations. Their last hurrah came after twelve thousand years ago, when the great traditions of painting and engraving faded away after more than twenty thousand years (figure 12.3).

We can watch the changes at a Spanish cave with twenty-nine occupation levels used by people who may well have sung and danced at Altamira. The small La Riera cave lies fifteen miles (twenty-five kilometers) west of the great shrine.[5] Twenty-three thousand years ago, ibex and red deer hunters visited La Riera, which overlooked a treeless landscape. Lawrence Straus, one of the excavators of the cave, believes that netlike symbols painted on the cave walls show the stout nets the hunters may have used to ambush mountain goats. By seventeen thousand years ago, the landscape had changed dramatically. Birch and pine trees now masked the cave entrance. People made greater use of the Atlantic shore, where they collected large numbers of limpets, urchins, and other shellfish, as well as spearing sea bream from the rocks. But

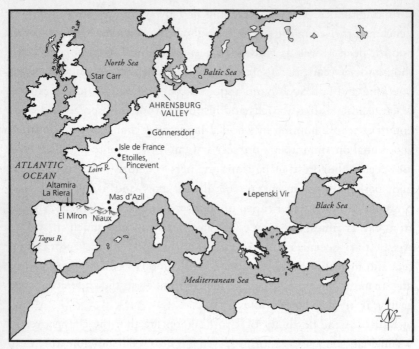

Figure 12.3 *Map showing sites mentioned in chapter 12.*

hunting continued, more often as an individual quest for deer and birds using bows and stone-tipped arrows. By now, the paintings and rituals at Altamira may have become irrelevant. Visitors to La Riera now relied heavily on wild seeds and tubers for much of their diet.

Over the next seven thousand years, the Atlantic rose rapidly, until the coast was little more than a mile (two kilometers) from the cave. Again, the inhabitants of La Riera adapted, this time by eating ever more limpets and other mollusks. A large midden formed inside the cave, an odiferous mass of rotting animal and fish bones and discarded shells. (A midden is an accumulation of human occupation debris.) We can imagine women dressed in skin blouses and pants pounding limpets by the dozen off the rocks with crude stone picks. Hefting hundreds of mollusks back to the cave in red deer skins, they would extract the meat and throw the empty shells on the rotting pile behind them. The size of the limpets in the midden became smaller and smaller over the generations as local populations rose and collecting along the shore intensified. Eventually, the visitors moved elsewhere: the midden was too large for the cave to offer comfortable shelter.

Broadly similar changes unfolded elsewhere, for the Pyrenees foothills and Cantabria, with their numerous caves, were a favored area for human settlement. Mas d'Azil, in Ariège, is a great tunnel where a river flows through a limestone cliff. Here people camped on the banks, close to the entrances of painted and engraved caverns, carving some of the finest examples of Cro-Magnon art to come down to us: horses, ibex, and waterfowl. Mas d'Azil's deposits are full of exotic seashells and marine fish bones, brought by people who came to the caverns from afar. This was an important ceremonial center during the final gasp of the Ice Age, and it remained a sporadic camping place long after the glory days of the artists. Eleven thousand years ago, no one had any interest in the nearby painted walls. They camped by the Arize River for the fishing. Like their La Riera contemporaries, they speared fish, this time with flat antler harpoons, which have none of the sophistication of earlier fishing equipment (figure 12.4).

The uppermost levels of the Mas d'Azil camps, on the opposite bank from the caverns, yield not only flat antler harpoons but also more than fifteen hundred painted pebbles, adorned with dots and other signs. French archaeologist Claude Courand studied both these specimens and five hundred others from sites elsewhere in France, Italy, and Spain. He identified sixteen signs and tried in vain to decipher the meanings of dot combinations between one and four. Were these numbers or, as Courand suggests, a record of lunar phases? We will never know, but they are certainly something very different from the cave paintings in the caverns.[6]

Cro-Magnons visited El Mirón cave, in the cliffs of Monte Pando, in the outer ranges of Spain's Cordillera, west of La Riera, over a much longer period, from as long ago as forty-one thousand years ago until nine hundred years before the present (figure 12.5). Neanderthals and early Cro-Magnon visitors stayed here. So did Magdalenians and later hunters. Human occupation expanded dramatically at several sites in the area after the Last Glacial Maximum, with intensive Magdalenian settlement at El Mirón between seventeen thousand and twelve thousand

Figure 12.4 *Azilian harpoon.*

years ago, during a period of major climatic fluctuations and frequent warming. The visitors hunted mainly ibex in the nearby uplands and sometimes more-lowland animals like red deer, fished for salmon and trout in the rivers, and brought in mollusks from the Bay of Biscay shore, a good day's walk away. After twelve thousand years ago, forests expanded and woodland species such as roe deer, boar, and chamois became favored prey. There was constant movement between the seashore and the mountainous interior, a pattern typical of much of Cantabria for thousands of years.[7]

The profound changes in daily life documented at La Riera, Mas d'Azil, El Mirón, and elsewhere reflect thousands of years of gradual adjustment to a new world, to very different landscapes. Similar, albeit diverse, changes took hold all over southern Europe during the same

Figure 12.5 *El Mirón cave (Ramales de la Victoria, Cantabria, Spain) in the Cantabrian Cordillera, between Bilbao and Santander, excavated by L. G. Straus and M. R.Gonzalez Morales since 1996, with a cultural sequence from the late Mousterian through the early Bronze Age, and dated by 65 14C assays ranging from 41,000 B.P. to A.D. 1400 (traces of the Middle Ages on the surface). Photo by L. G. Straus. Copyright Lawrence G. Straus.*

millennia. For all the adjustments, the basic realities of Cro-Magnon life remained unchanged, even if people no longer communed with the spiritual representatives of their prey in dark caverns. There may have been more people living in smaller hunting territories, but the need for sharing information and for cooperation became, if anything, greater in changed, more wooded environments, where game was harder to spot, where survival often depended on nut harvests and edible seeds, and where such seasonal events as salmon runs provided critical reservoirs of stored food for the winter.

Eight hundred generations of Cro-Magnons had gained power from the supernatural forces of animal spirits residing behind painted cave walls, adorned with great beasts and with more-prosaic quarry. Now the prey had changed, life was lived more in the open, and the chase was more solitary, so the relationship between the hunter and the hunted had changed. If historic hunting peoples are any guide, new generations of Cro-Magnons still treated their prey as sentient beings, to whom they talked and with whom they communed, but the ways in which they did so may have changed with the shifting realities of the chase, with an entirely new form of existence that no longer depended on great beasts and communal hunts. At the same time, the canvas had changed, too, from antler, bone, and rock faces to wood and other perishable materials.

THE ENTIRE HISTORY of the Cro-Magnons was one of movement, of countless ebbs and flows across ever-changing landscapes. As the Ice Age ended and temperatures warmed, some groups stayed where they were and adapted effortlessly to new ways of obtaining food, becoming plant eaters and fishers as much as hunters. Others followed their traditional quarry northward into the still-open fastnesses of the north, to the realm of the fast-retreating ice sheets and the meltwater lakes that were to form the Baltic Sea thousands of years later. They moved northward from more-sheltered areas, from refuges like the tributaries of the Loire River, which fingered northward into the tundra.

After about 17,000 years ago, small numbers of Cro-Magnons moved into uninhabited areas, starting with the Upper Rhine Valley, then, by 16,500 years ago, heading downstream and into Belgium and southern Germany. They were following horses and reindeer herds moving into what had been polar desert. Well over a hundred state-of-the art radio-carbon dates document these moves and tell us that the first hunting bands moved north at a rate of about a sixth of a mile (one quarter of a kilometer) a year.[8] By 16,000 years ago, Cro-Magnons had colonized northern France and northern Germany, as well as Denmark. Under two thousand years later, a few bands had hunted their way into Britain, which was, of course, still part of the continent. The warmer temperatures of the Bölling oscillation accelerated the push.

Like newcomers everywhere, the first settlers left few signs of their arrival, just scatters of stone artifacts that mark transitory hunting camps occupied during the summer. During the cold months, the hunters would return to winter bases in more sheltered locations. Radiocarbon dates say that this initial phase of settlement lasted about five centuries, perhaps a time of familiarization, of obtaining intelligence about game, water supplies, sources of toolmaking stone, and so on. These were times of constant temperature shifts, of drastically changing circumstances on the open plains, which may have made intelligence out of date from one year to the next and delayed the establishment of base camps in the north. Precisely the same general process may have unfolded thousands of years earlier when the first Cro-Magnons spread into Europe.

After about twenty generations, the settlement pattern changed. Hunting bands established permanent territories in previously inaccessible landscapes, which now supported humans and the animals that fed them. The warmer temperatures of the Bölling were a time of population growth, perhaps reduced infant mortality, and certainly great mobility. The new lands stretched to the far horizon, perhaps not inviting, but certainly offering food and opportunity to people whose ancestors had lived through long, harsh winters and often gone hungry. There were still bitterly cold winters and occasional hunger, but the built-in flexibility of hunter-gatherer life meant that Cro-Magnons, like the prey they sought, moved northward as part of the endless cadence of now better-watered lands.

By fifteen thousand years ago, Magdalenian groups hunted on the Île-de-France and elsewhere in the Paris Basin. One site lies at Pincevent, near the confluence of the Oise and Seine rivers, where some Magdalenian reindeer hunters camped during the Bölling. An estimated five families pitched a cluster of reindeer-hide tents, each with its own hearth. The inhabitants returned to camp with partially butchered reindeer carcasses and numerous antlers, then cut up the bodies and divided them between the families. In a remarkable piece of research, the American archaeologist John Enloe was able to match up the right forelimb of a reindeer at one hearth with the left forelimb of the same beast at another. This is one of the first documented examples of food sharing in history, but a routine practice in Cro-Magnon societies, where sharing, like cooperation, was an integral part of daily life and managing risk.[9]

The Paris Basin had other attractions as well, notably fine-quality flint, easily extracted in large nodules from the chalk and limestone outcrops in local river valleys. At Etiolles, in the Seine Valley, about twenty-five miles (forty kilometers) north of Pincevent, archaeologist Nicole Pigeot spent months rejoining flakes and cores from five clusters of flint debris from about 14,500 years ago. The flint nodules that displayed the fewest errors lay closest to the hearths, where, presumably, the most skilled stoneworkers sat. Progressively less-skilled practitioners worked at increasing distances from the warmth. Those at the edge were very unskilled indeed. Etiolles was so close to the flint outcrops that there was plenty of stone for practice; in most places, really good-quality flint was used conservatively, much of it by adept stoneworkers.[10]

The landscapes of the north were productive hunting grounds, where a scattered population preyed on horses and reindeer. Life here would never have been easy, dependent as it was on reindeer herds, whose numbers and migrations changed abruptly from one year to the next. As always, the keys to survival were intelligence and mobility, which meant regular contacts between bands living long distances apart, and every group must have covered enormous distances every year. Of course, the only way we have of studying this is examining the exotic objects that appear in archaeological sites of the day. This time, however, we have the smoking gun. Shells found in southern Belgian caves come from geological strata near Paris and in the Loire Valley, 93 miles (150 kilometers) and 217 miles (350 kilometers) away, respectively.

As always, the staple was *Rangifer tarandus*, especially in the autumn when the beasts were in prime condition. The hunts followed a familiar pattern. For example 14,600 years ago, a group of hunters ambushed a herd of north-migrating reindeer in autumn as they passed between two small glacial lakes in northern Germany's Ahrensburg Valley, south of Hamburg. Alfred Rust excavated the Meiendorf site before World War II. He found thousands of reindeer bones in the muddy deposits of the valley floor and also the stone projectile points that had speared the ambushed herd. Many of the trapped and wounded reindeer had stampeded into the lake. While the hunters had killed and butchered the animals on dry land, they had left the unreachable carcasses floating on the water to sink. Meiendorf is a striking example of the effectiveness of spear throwers, which helped a relatively small group of hunters dispatch well over a dozen reindeer in a few minutes.[11]

Figure 12.6 *Gonnersdorf figures. Interfoto Pressebildagentur/Alamy.*

The rolling terrain of west-central Europe was still quite open land-scape during the Bölling. Here, too, hunting groups traveled long distances, exchanging flint, amber, jet, and other commodities over sixty-two miles (one hundred kilometers) from their sources. Family groups would come together to pursue horses in river valleys during autumn, when the beasts were fat. Two locations, Gönnersdorf and Andernach, lie opposite one another across the Middle Rhine and were in use between about 15,000 and 12,500 years ago. The inhabitants of the former lived in substantial winter huts supported by thick wooden posts and covered with turf and hides. As many as a hundred people may have gathered in the camp over the winter, convenient to the horse herds grazing by the river.

Gönnersdorf is famous for its engravings, inscribed on over fifteen hundreds slabs adorned with animals and women. The artists depicted horses and mammoths with consummate naturalism, down to the details of eyes and tails. They also occasionally drew birds, seals, woolly rhinoceroses, and lions. The engraved figures of the women lack the naturalism of the animals. Sometimes the artists engraved a complete female body, at other times abstract images of back and buttocks, even groups of women in lines, giving the effect of dancing (figure 12.6). Similar abstractions of the female body appear at other sites, but never so plentifully as at Gönnersdorf. Why women? Like the Venus figurines of earlier times, they remain a Cro-Magnon enigma.[12]

EVEN DURING THE Bölling oscillation, conditions in the north were harsh, even if summers were longer and somewhat warmer, at times with temperatures approaching those of the present day. Then, suddenly, 12,900 years ago, temperatures plunged rapidly, perhaps in the course of a century or less, during a thousand-year cold snap, known as the Younger Dryas (named after a polar flower) that plunged Europe into near-glacial conditions once more. The sharp temperature drop may have resulted from a partial shutdown of the Gulf Stream in the North Atlantic caused by a massive inflow of glacial meltwater from the vast Laurentide ice sheet, in northern Canada. The woodlands retreated, the tundra expanded, and glaciers advanced for a millennium. Nevertheless, a sparse population of hunting bands used base camps in large river valleys like that of the River Elbe. In spring and, especially, late summer and fall, they would move out into small tunnel valleys like the Ahrensburg, where they would prey on migrating herds.

The Stellmoor Hügel is a small hill protruding into the Ahrensburg Valley, offering a superb view of the surrounding terrain, which was at the time open tundra.[13] A long, shallow lake covered most of the valley floor, and hunters ambushed reindeer here on many occasions around 10,800 B.C. Alfred Rust dug into the waterlogged lake deposits during the 1930s and recovered not only the victims of the hunt, but the wooden and stone-tipped arrows used to shoot them. Some of the reindeer bones bear wounds inflicted by flint points, others lesions caused by wooden arrows. Bodil Bratlund has studied the wounds, and his research shows that the hunters shot at the ambushed beasts from slightly in front, from the side, and from behind. Some of the shots also came from behind and above, almost invariably hitting the neck and shoulders, as if the animals were swimming in the lake. Most of the wounds did not immediately kill or lame the reindeer, only about a third of the shots being fatal. All the wounds were fresh, so the hunters probably acted in concert, one wounding a beast, another eventually killing it.

Once the killing was over, the parka-clad hunters butchered the victims on the lakeshore. Of the fifty or so reindeer killed at the site in multiple episodes, the forelegs and backs of only about twenty were consumed locally. The butchers removed antlers, hides, and sinews from all the partially butchered animals, then left everything else. We can imagine them working quickly and efficiently, cutting off the antlers, examining them closely, then breaking them up, keeping only the choice lengths for toolmaking: They cut through the hides at each ankle and peeled them rapidly off the carcasses, rolling them up for transport back to camp. Sometimes they set aside the fore- and hind limbs for meat, but they spend more time working over the neck and backbone of each beast, teasing out the sinews that have so many potential uses. The surviving Stellmoor bones represent a tiny fraction of the number of animals killed there during generations of autumn hunts. The best estimate for multiple incidents is about 1,350 beasts of all ages. The hunts may have taken place in September or October, when local bands moved to strategic locations after hunting geese and swans in the summer at places nearby. During the remainder of the year, people hunted prey opportunistically, including horses.

Stellmoor is the only site in the north where we are certain that bows and arrows were used. Only pine arrows came from the excavations, which is interesting, because wood must have been in short supply on the treeless open landscape. While bows and arrows, with their superior range, may have been a major factor in the settlement of the open country in the north, the hunting territories of those who used them must have included places where they could obtain wood for their weapons.

UNTIL NOW TEMPERATURES had been generally cold, but warming was now under way. The landscape of northwestern Europe was changing profoundly as the northern ice sheets retreated and sea levels rose. The North Sea was a low-lying, marshy expanse, with the forerunner of the Elbe running northward into the deep waters of the Norwegian Trench. The Rhine, the Seine, and the Thames flowed into a wide estuary between Britain and France that is now the English Channel. When warmer temperatures returned after the Younger Dryas, about eleven thousand years ago, human settlement flowed to the north, into a dynamic, rapidly changing world of shallow estuaries, mudflats, and sand banks with staggering biodiversity. Shallow inshore waters and sheltered inlets provided fish, bird life abounded, shellfish of all kinds were plentiful, and plant foods such as edible seaweed were there for the collecting. These were very different environments from those of the tundra; family bands could settle in one place for months at a time, even permanently.

Unfortunately, rising sea levels have obliterated the traces of many of the communities that once flourished along the low-lying coastlines of the north. Artifacts dredged up from the North Sea confirm that hunting bands lived on what was then dry land, but we have only snapshots of the very earliest settlers. One group visited the shores of a glacial lake at Star Carr, in northeast England, 10,500 years ago, camping there in spring and early summer as they hunted deer in the nearby birch woods. On several occasions, they burned the reeds that fronted the water, to get a better view and also to launch their dugout canoes, of which only a paddle survives today.[14]

The Star Carr people lived on slightly higher ground, but their distant cultural relatives along the Baltic coastline could never be sure if their camps would be above water for long. Estuaries flooded in a few days—disappearing forever. Quiet creeks became shallow lakes; sandy beaches could vanish overnight. By 10,000 B.C., oak forests pressed close to shore, but a few days of high water caused by a storm could flood their roots with saltwater, and even large trees would perish. A sharper contrast to the tundra it is hard to imagine, but the environment was a veritable Garden of Eden to those who fished its waters . . . [15]

The weathered dugout canoe glides silently across the shallow bay on a calm summer night, the man in the bow, his wife paddling at the stern. They wear light bird-skin anoraks and pants, for it is cool on the water. Her paddle barely makes a ripple in the moonlight as they pause by each of their wicker traps. The man reaches over and grasps the narrow end of one of the tunnel-like baskets. He lifts it quickly, the contents wriggling silver against the dark water. He tips it into the canoe. A stream of eels cascades between his feet. A short time after-

ward, husband and wife return to shore with a full load, their canoe
alive with the slithery catch. They beach the dugout on the shingle below the camp and then, grabbing eel after eel, kill each with a quick
blow and gut it with a sharp flint knife. As the moon sets, they hang the
eels above their smoky hearth, where they will stay until they are as dry
as sticks . . .

Dugout canoes, wooden fishing spears, nets, fish traps—the inventory of simple equipment used by the coastal fishers has survived in
waterlogged settlements on the Baltic shore, some of them in shallow
water. Two weapons were all-important: the barbed fish spear, often
with double or triple heads, and the bow and arrow, used not only
against deer and aurochs but also against birds in flight. The hunters
used stone-tipped arrows of many kinds, carried in quivers that
weighed less than a single spear and spear thrower . . .

The bowman crouches downwind, absolutely still in the reeds,
watching the migrating waterfowl come in to land on the shallow lake.
The weary birds are moving fast, only a short distance from the ground.
With effortless, unhurried ease, the young man aims and shoots. A bird
falls dead near his feet, then another, and another. A few fall into the
water and float motionless, carried toward the shore by the wind. He
wounds two birds, which flutter helplessly in the reeds. Two quick blows
from a wooden club, and he adds them to the catch. His sister paddles
out in a small dugout and picks up the dead quarry floating out of
reach . . .

No spear could bag a bird, but an archer of reasonable skill could
shoot a goose out of the sky, or at least stun it with a fast-moving arrow,
then club it when it fell to ground. The effectiveness of the arrow depended on its tiny, delicately fabricated point, with a tip so sharp that it
could penetrate fur and a tough hide.

Food supplies were so abundant along northern coasts that populations rose steadily, so much so that larger numbers of people crowded
into already intensely exploited hunting territories. The territories
shrank; competition for shellfish and fishing grounds, for plants and
hunting grounds, intensified. Inevitably, such competition erupted into
violence after nine thousand years ago. Cemeteries in some coastal

Microliths—another Swiss Army Knife

If there is one thing about Cro-Magnon tool kits that everyone agrees upon, it is that they got progressively smaller, especially after the Last Glacial Maximum about 18,000 years ago. This was a natural development of the Swiss Army Knife technology, in which blade cores became increasingly smaller and more efficient. The smaller stone tools were mainly, but not invariably, sharp-edged blades with steeply trimmed backs that were almost certainly stone barbs set into antler, bone, and wooden spearheads. Some of them were so small that some experts believe Cro-Magnons were using the bow and arrow earlier than eighteen thousand years ago, but there is no definite proof. Most likely, they used light spears propelled by spear throwers, ideal weapons for hunting horses, reindeer, and other herd animals pursued in the open.

We know more or less for certain that the bow and arrow first appeared south of the Pyrenees sometime in the very late Ice Age, but all we know of the new weapon comes from small tanged stone points, which most likely, because of their lightness, served as arrow barbs. The first absolutely certain use of bows and arrows was at the Stellmoor reindeer kill site of 10,800 B.C., with its pinewood arrows, by which time the new weaponry may have been in increasingly widespread use. Both bows and arrows require good-quality timber with a straight grain, which was rare indeed in the open landscapes of pre–Younger Dryas times. For this reason, it seems likely that bows came into their own as forests spread across Europe after ten thousand years ago. With their enhanced range and high velocity, bows were ideal for forest hunting and, above all, use against flying birds, where speed, accuracy, and good penetrating power was essential. While many of the Stellmoor arrows were made entirely of wood, others had small stone tips, which were to become the projectile point of choice in later millennia. Once again, the Swiss Army Knife came into use, with the development of the microlith.

Microliths (Greek for "small stone") are small bladelets punched off from a blade core, then turned into increasingly geometric artifacts of various sizes and shapes. A punch removed the blade, which was thicker at the end where it was struck off. The stoneworker now shaped the tip and sides of the bladelet to produce the barb or other tool he or she required. Having fabricated, say, a microlith with one side blunted for mounting and the other sharp, the stoneworker then notched the blade and snapped it on one side. Presto, the thicker end could be discarded, and the microlith could be mounted on a weapon.

Microliths came in many shapes: backed bladelets, surely the prototypes from earlier times for the microlith; crescents; triangles; trapeze shapes; and many others. They served as barbs for spears and arrows, made fine awls for tailoring, and could be used as parts of woodworking tools. Just like the Gravette point of earlier times, they were multipurpose artifacts, made in a few moments and used for all kinds of tasks (figure 12.7).

By any standards, microliths made for lethal hunting weapons. Back in the early years of the twentieth century, the University of California researcher Saxon Pope went hunting with Ishi, the last Yahi Indian in northern California to practice ancient ways, using only traditional weaponry. Pope noticed that stone points are more effective than steel-tipped arrows against deer and birds.[16] They are sharper. A stone point enters the quarry obliquely, cuts the skin, and does serious damage to the organs it encounters. Add a second armature like a barb, and the arrow inflicts a much larger wound. The most effective barbs of all form lateral cutting edges, especially effective when several of them are mounted on the same shaft.

Microliths were the last refinement of Cro-Magnon blade technology, which depended on the Swiss Army Knife principle taken to minute ends. The final manifestation of the technology appeared about ten thousand years ago, when hunters all over Europe began notching their blades on both sides to form trapeze-shaped microliths

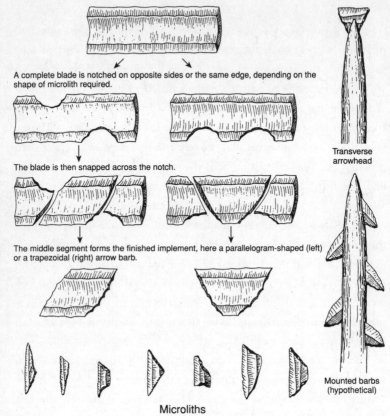

A complete blade is notched on opposite sides or the same edge, depending on the shape of microlith required.

The blade is then snapped across the notch.

The middle segment forms the finished implement, here a parallelogram-shaped (left) or a trapezoidal (right) arrow barb.

Transverse arrowhead

Mounted barbs (hypothetical)

Microliths

Figure 12.7 *Microlith technology and microliths.*

that could be mounted transversely at the tip of the arrow. Given that people all over the continent adopted the trapeze at about the same time, it must have made for a devastatingly effective weapon, and it would have worked well against birds in flight.

settlements contain victims who had been clubbed to death, others with arrow points in their bones, wounds inflicted in violent, face-to-face fighting at close quarters as people battled over turf.

OVER MUCH OF Europe, the descendants of earlier Cro-Magnon societies lived in dense woodland, relieved by occasional clearings, lakes,
and marshes. No hunting band numbered more than a few families,
who subsisted in part on forest game like deer and the occasional boar
or aurochs. Their staples were small animals like rabbits, rodents, and
birds shot on the wing. Above all, they subsisted on wild plant foods:
tubers and nuts, fruit and edible grasses.

If you walked in the forest of that time, you would enter a dark world
of tall trees and often-dense undergrowth, where there were few signs
of people. You could rest assured, however, that silent observers were
watching you from only a short distance away. You might smell acrid
wood smoke from a campfire, perhaps hear a barking dog, but the telltale sign of a nearby settlement would be a monotonous *scrape,
scrape*—a milling stone rubbing wild seeds on a grinder. This was a very
different world from that of the tundra, or of the hunter-gatherers of
northern Spain, one where life was defined by plants, where skin-clad
women spent hours each day grinding and preparing food. You can
imagine such a woman kneeling in front of a flat stone, slightly rough-
ened in the middle. The seeds congregate under the milling stone,
slowly turning into fine meal under expert hands. The woman adds
more seed, turns the milling stone to even the grind, tips the stone to cir-
culate the grain, a process that is as much a part of her life as sleeping or
tending a fire.

The forest world was different, but the realities of survival never
changed: the months of plenty in summer and fall; the long, cold
months of winter and early spring, when the world seemed to be asleep,
often the time of hunger and dying.

SOME GROUPS LIVED in wonderful places. On the other side of Europe
from the Baltic, the Danube River meandered across the undulating
forest landscape, then entered the steep defile of the Iron Gate on the
boundary between Serbia and Romania, which cuts through the
Carpathian Mountains. During the last glaciation, small groups of
hunters preyed on ibex and caught salmon here, in one of those rare

places where forests survived during the Last Glacial Maximum. With the warming, tree cover thickened, rainfall increased, and food became more abundant. At first people wintered here, then they stayed longer and longer. By eighty-five hundred years ago, they dwelled year-round at a site named Lepenski Vir, in the heart of the Gate.[17] Here great sturgeon migrated upstream in spring, some of them up to twenty-nine feet (nine meters) long. Here, too, game abounded; plant foods were nearby for the taking. The inhabitants built trapeze-shaped dwellings with carefully completed floors and central hearths, with paths between them that led to an open space in the middle of the community by the river. The river was the catalyst not only for settlement but also for a complex ritual life, commemorated by stone sculptures found throughout the village. Most are boulders carved into portraits that seem, sometimes, to combine elements of humans and fish. These portraits lie in the foundations of the houses, as if they were meant to link the owners to the great river, with its life-giving powers, that flowed at their doorsteps.

Nearly six thousand years have passed since the Ice Age, yet the descendants of the reindeer hunters still live by hunting and foraging, in a very different world. Their lives have changed beyond recognition, and are more comfortable, perhaps, but human existence is still harsh and unforgiving, even for people living in such favored locations as Lepenski Vir, where the bones of the dead display the telltale signs of occasional malnutrition and dietary stress. People survive because they know their environments intimately, because they depend on others, and also because of the rich ceremonial life and intricate ritual beliefs that link them to the animals they hunt and to the complex forces of the natural world. In that respect, nothing has changed, for one of the most powerful legacies of the Cro-Magnons was their ability to integrate the intangible and material worlds into a single human existence.

LIFE WAS ABOUT to change profoundly. Unbeknownst to the Europeans, the thousand years of the Younger Dryas had brought savage droughts to the Near East after a long period of wetter conditions.[18] Forests retreated; the yields of autumn nut harvests plummeted; people fell back on edible seeds and other less desirable foods. As natural stands of grasses withered, those who harvested them planted the seeds to supplement the yield, a logical way to preserve their traditional means of obtaining food. Agriculture was never a dramatic invention, but within a surprisingly few generations, the people of the Near East and southeastern Turkey were entirely dependent on farming. When wetter conditions returned at the end of the Younger Dryas, the new economies spread like wildfire across Anatolia and into southeast Europe, where they were well established before eight thousand years ago. Five centuries later, farming communities appeared on the Hungarian Plain. Their descendants spread rapidly across Central Europe and into Ukraine.

Within a few centuries, the descendants of the first farming colonists—known to archaeologists as the Linearbandkeramik (LBK) people, after their distinctive bag-shaped clay pots—had expanded across Europe in a northwesterly direction, covering the ground at a rate of over fifteen miles (twenty-five kilometers) per generation.[19] By seven thousand years ago, they had settled in the Low Countries. Each village cleared small patches of light, fertile soil, much of it the weathered loess deposited by the glacial winds of the Ice Age. The inhabitants would settle near reliable water supplies, clear fields for growing barley and wheat, and then leapfrog onto new tracts of farming land some distance away when the original soil became exhausted. They dwelled in timber and wattle-and-daub longhouses, up to thirty-nine feet (twelve meters) long and divided into three sections, one for living, one for cooking and eating, and one for storage (figure 12.8). The small farming communities were self-sustaining, grazing their cattle in the surrounding woodland, their pigs on the undergrowth. These were family-based villages, some no more than a single longhouse, others much larger; the dwellings were abandoned when the members of the household died.

Everywhere these farming communities settled, dense forest pressed on river valleys and farm-land, a dark, mysterious place. Within the deep shadows, trunks and roots of fallen trees rotted on the forest floor, pathless except for the occasional game trail. Carpets of bright green

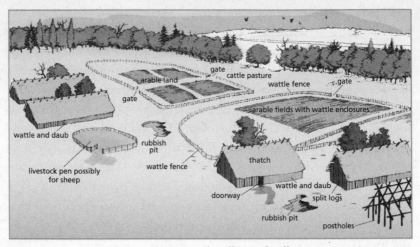

Figure 12.8 *Linearbandkeramik village.*

moss lay underfoot, surrounding ponds and deep bogs of waterlogged vegetation. A clearing here and there let sunlight between the tangled trees, revealing grazing bison, deer, and elks, who would silently vanish when a hunter appeared. When the farmers arrived, the primordial forest stretched to the far horizon, pristine and undisturbed except when the indigenous population fired the undergrowth to attract game to feed on new shoots. Only a few thousand hunters dwelled among the trees. They were elusive, cautious people, armed with bows and arrows and an intimate knowledge of woodland plants: bog cranberries, mushrooms, wild garlic. Like the Neanderthals thousands of years earlier, they were silent people. We can imagine them, as the seasons pass, watching from the shade, studying men and women clearing the land, staying carefully upwind of the acrid smoke as the newcomers set fire to dry grass and undergrowth in the fall. The indigenous inhabitants track cattle and pigs foraging at forest's edge and melt quietly away as the farmers harvest acorns from the great oaks on the edge of the valley.

At times, hunter and farmer would meet, at first cautiously, perhaps exchanging honey for grain, emmer meal for elk hides. Sometimes the two would fight and arrows would fly, as a new village encroached into ancient hunting territory. But contacts became routine over the generations, perhaps somewhat like the contacts between southern Africa's

Lala farmers and *utunuta mafumo* described in chapter 1. And, eventually and inexorably, some of the hunters served as cattle herders or stole some beasts and started off on their own. Over time, more and more groups became farmers, at least part-time, and the ancient foraging lifeway began to pass into history. Farmers married women from the forest; their children never became hunters but founded their own longhouse communities some distance away. For centuries, there were hunters at the periphery, as the traditions whose roots went back to the very first Cro-Magnons persisted, but eventually they vanished.

As agricultural communities spread across central and western Europe, other farming villages appeared along the Mediterranean coast, many of them probably founded by hunters-turned-farmers.[20] Significant numbers of the surviving hunting groups appear to have based themselves in caves and only adopted part of farming culture, such as sheep raising, pottery, or cereal agriculture, to fill gaps in the seasonal round. Flourishing communities of hunter-gatherers may have survived for centuries alongside farmers in enclaves where fish, shellfish, and plant foods were abundant. One such location would have been the Tagus and Sado river valleys of central Portugal, where huge shell middens that also served as burial grounds remained in use until about 5,000 B.C.[21]

A thousand years later, the changeover was complete. Europe had become a continent of farmers, except in the far north, where hunting survived until about three thousand years ago, when the ancient hunter-gatherer traditions finally expired.

WHAT, THEN, WAS the legacy of the Cro-Magnons? The answer may come from genetics.[22] The genes of Europeans alive today, argues the geneticist Luca Cavalli-Sforza, show a gradient from east to northwest, the legacy of farming people, who arrived from Greece, settled in the Balkans, and then spread northwest across Europe. He believes that this genetic "wave of advance" coincided with the spread of the first farmers. Brian Sykes of Oxford University and his colleagues disagree. They use our old friend mitochondrial DNA, which has a faster mutation rate and thus gives you, at least in theory, a more accurate picture of history in short chunks of time. Sykes examined mtDNA from 821 individuals across Europe and found six clear lineage groups. Thus, Europeans were much more genetically diverse than one might expect from Cavalli-Sforza's wave theory. Only one of these genetic groups was late enough on the molecular clock to be linked to the arrival of farmers from western Asia. Not only that, but the same group boasted genetic markers that were of western Asian origin. Furthermore, the distribution of this particular group in Europe matched that of the two farmer population movements, the earlier one by LBK people and the later one along the Mediterranean. Only 15 percent of the lineages within the six groups shared this genetic makeup. All other lineages were much older: dating to between twenty-three thousand and fifty thousand years before the present. That is, 85 percent of the mtDNA lineages in Europe were already present long before farming arrived and could be attributed to the indigenous populations that we loosely call Cro-Magnons. The genetic impact of the farmers was effectively negligible. More recent calculations by both Cavalli-Sforza and Sykes have increased the percentages of immigrants sides to 28 percent and 20 percent respectively, so the dispute is fundamentally resolved.

Thus, no one can doubt that the descendants of the Cro-Magnons, the most ancient fully modern Europeans of all, played an active role in the development of farming in Europe. And in so doing, they sounded the death knell of a way of life that had flourished for more than thirty-seven thousand years. The genes of the Cro-Magnons are still dominant among modern Europeans today. My DNA tells me that genetically I'm one of them, and I'm proud of it.

Acknowledgments

My fascination with the Cro-Magnons began when I was a freshman at Pembroke College, Cambridge. Miles Burkitt, who had copied cave paintings and excavated with his mentor, the Abbé Henri Breuil, before World War I, was never a particularly distinguished archaeologist, but he had the gifts of enthusiasm and storytelling. He gave me a lifelong passion for archaeology and for the Stone Age. Burkitt introduced dozens of would-be archaeologists to the past; his celebrated admonitions to "never let the sun set on an unmarked implement" are as relevant today as they were a half century ago. In the final analysis, he inspired this book, as did Charles McBurney and Eric Higgs, who trained me in Stone Age archaeology.

Spencer Wells, geneticist extraordinaire, encouraged me strongly to write *Cro-Magnon* after we visited the Les Eyzies National Museum of Prehistory together. I am grateful to him for his enthusiasm. This book is based on years of site visits, many hours in collection rooms, and extensive travel. Such activities alone have been a challenge, but they pale into insignificance alongside the literature, which presented a unique ordeal even by the standards of archaeology, a field remarkable for its contradictory and obscure publications.

A large number of colleagues and specialists have contributed to my better understanding of the subject. They have answered questions, allowed me to listen in on discussions, steered me to little-known sources, and challenged my reasoning. It's impossible to name everyone from a list of indebtedness that goes back decades, and I hope that those who are not mentioned will forgive me and accept a collective heartfelt thanks. Special thanks to Stanley Ambrose, Paul Bahn, Christopher

Chippindale, Clive Gamble, David Lewis-Williams, Paul Mellars, George Michaels, John Shea, Mary Stiner, Lawrence Straus, and Chris Stringer, all of whom have assisted me in one way or another over the years. John Hoffecker read the entire manuscript with trenchant care and saved me from many sins. So did Aaron Elkins, mystery writer and anthropologist, who gave me the benefit of his unique dual experience. I am grateful to all those who gave permission for the reproduction of illustrations. Every effort has been made to contact copyright holders. Any queries should be addressed to the author.

As always, Shelly Lowenkopf was at my side throughout the writing of the book, encouraging, suggesting, and offering insights as only a non-archaeologist and fellow writer can do. Steve Brown performed his usual feats of magic with the drawings; Francelle Carapetyan was a tower of strength in obtaining the photographs, which was sometimes a frustrating challenge.

My debt to my colleagues at Bloomsbury Press is enormous, notably to Peter Ginna and Pete Beatty, whose perceptions and editorial skills transformed the manuscript. I value their friendship and belief in my work more than I can say. Michael O'Connor and Nate Knaebel worked miracles during production. Susan Rabiner, agent extraordinaire, was behind me all the way. Lastly, my usual word of thanks to Lesley and Ana, as well as our various beasts, all of whom are always there for me. I think that our cats are probably world champion keyboard sitters. They can claim no credit for these pages!

Brian Fagan
Santa Barbara, California

Notes

An enormous amount literature in many languages surrounds the Neanderthals and the Cro-Magnons. Much of it appears in obscure archaeological journals and in edited volumes resulting from conferences that
seem to proliferate like rabbits almost monthly. Most of the publications
are, of course, highly specialized and of little relevance to these pages. The
references that appear below provide a cross section of the literature as of
late 2008 and contain useful bibliographies for those wishing to probe
deeper. For obvious reasons, I have tended to list sources in English. Fortunately, much of the most important literature is in this language, now
widely used internationally for science. The interested reader will find key
citations to foreign-language contributions in specialized bibliographies.

Chapter 1: Momentous Encounters

1. Many specialists consider the term *Cro-Magnon* inappropriate. They
prefer to refer to *modern humans* or *anatomically modern humans (AMH)*. I have
unashamedly used the term *Cro-Magnon*, largely because it is widely known.
Its use here is purely generic and implies no specific time frame or cultural affiliation. I consider the usage appropriate for a popular work. After all, the first sci-
entifically identified modern humans in Europe came from the Cro-Magnon
rock shelter.

2. For a general account of the Neanderthals and their ancestors, see Chris
Stringer and Peter Andrews, *The Complete Book of Human Evolution* (London:
Thames and Hudson, 2005).

3. Louis Lartet (1840–1899) was one of the pioneers of late Ice Age archaeology. His father, a lawyer and paleontologist, was the first archaeologist to work
seriously at Les Eyzies.

4. For a description of Lartet and Christy's excavations, see Édouard Lartet and Henry Christy, *Reliquiae aquitanicae* (London: Williams and Norgate, 1875). The Oxford geologist William Sollas compared the Cro-Magnons to Eskimos in his *Ancient Hunters and Their Modern Representatives* (London: Macmillan, 1911).

5. The issues are well summarized in Richard Klein and Blake Edgar, *The Dawn of Human Culture* (New York: Wiley, 2002). See also Spencer Wells, *The Journey of Man: A Genetic Odyssey* (Princeton, NJ: Princeton University Press, 2002).

6. R. E. Green et al., "A Complete Neanderthal Mitochondrial Genome Sequence Determined by High-Throughput Sequencing," *Cell* 134, no. 3 (2008): 416–26.

7. I follow common usage here. *Eskimo* refers to U.S. arctic peoples, *Inuit* to Canadian. It is worth noting that the earliest known needles in western Europe date to about eighteen thousand years ago or later, but I am sure that wooden ones were in use earlier. How would the tall Cro-Magnons have survived the rigors of the Last Glacial Maximum without tailored clothing?

8. J. D. Clark, "A Note on the Pre-Bantu Inhabitants of Northern Rhodesia and Nyasaland," *South African Journal of Science* 47, no. 1 (1950): 80–85.

9. I am grateful to Sir David Attenborough for sharing this account with me.

10. The late Professor Eric Axelson kindly provided me with a translation of this Portuguese passage: *Cartas dos Vicereis do India* (Lisbon: Arquivo Nacional da Torre do Tombo, n.d.), 1–161.

Chapter 2: Neanderthal Ancestors

1. This passage is based on R. Dale Guthrie, *The Nature of Paleolithic Art* (Chicago: University of Chicago Press, 2005). Quote from 214.

2. A good basic description of *Homo ergaster* appears in Stringer and Andrews, *Human Evolution*, 132–38 (see chap. 1, n. 2).

3. Guthrie, *Nature*, 216.

4. Milutin Milankovitch, "Memories, Experiences and Perceptions from the Years 1909–1944," *Proceedings of the Serbian Academy of Sciences* 195 (1952): 1–322. (Original paper in Serbo-Croatian.)

5. For an excellent summary of Milankovitch's work, see John Imbrie and Katherine Palmer Imbrie, *Ice Ages: Solving the Mystery* (Cambridge, MA: Harvard University Press, 1979), chap. 8.

6. For a summary of Ice Age climatology and events, see Brian Fagan, ed., *The Complete Ice Age* (London: Thames and Hudson, 2009).

7. Leo Gabunia et al., "Earliest Pleistocene Hominid Cranial Remains from Dmanisi, Republic of Georgia: Taxonomy, Geological Setting and Age," *Science* 283, no. 5468 (2000): 1019–25.

8. Eudald Carbonell et al., "The First Hominin of Europe," *Nature* 452 (2007): 465–69. For a summary, see Stringer and Andrews, *Complete Book*, 144ff (see chap. 1, n. 2). Climate data is in Hugues-Alexandre Blain et al., "Long-Term Climate Record Inferred from Early-Middle Pleistocene Amphibian and Squamate Reptile Assemblages at the Gran Dolina Cave, Atapuerca, Spain," *Journal of Human Evolution* 56, no. 1 (2009): 55–65.

9. For Britain, see Christopher Stringer, *Homo Britannicus* (London: Alan Lane Science, 2006).

10. Naama Goren-Inbar et al., "Evidence of Hominin Control of Fire at Benot Ya'aqov, Israel," *Science* 304, no. 5671 (2004): 725–27.

11. Stringer and Andrews, *Complete Book*, 148–51. It should be noted that the Mauer jaw displays some Neanderthal features, to the point that some experts place European *heidelbergensis* fossils with the Neanderthals.

12. Ibid., 152–53, for the human remains. For the climatic data, Blain et al., "Long-Term Climate Record," 55–65.

13. Hartmut Thieme, "The Lower Palaeolithic Art of Hunting: The Case of Schöningen," in Clive Gamble and Martin Porr, eds., *The Hominid Individual in Context* (London: Routledge, 2005), 115–32.

14. Hartmut Thieme, "Lower Palaeolithic Hunting Spears from Germany," *Nature* 385 (1997): 807–10.

15. D. Mania, "The Earliest Occupation of Europe: The Elbe-Saale Region (Germany)," in W. Robroeks and T. van Koifschoten, eds., *The Earliest Occupation of Europe* (Leiden, Netherlands: Analecta Leidensia, 1995), 85–101.

16. F. Clark Howell, "Observations on the Earlier Phases of the European Lower Palaeolithic," *American Anthropologist* 68, no. 2 (1966): 111–40.

17. Mark Roberts and Simon Parfitt, *A Middle Pleistocene Hominid Site at Eartham Quarry, Boxgrove, West Sussex* (London: English Heritage, 1999).

Chapter 3: Neanderthals and Their World

1. For the discovery of the Neanderthals and changing ideas about them, see Clive Gamble and Chris Stringer, *In Search of the Neanderthals* (London: Thames and Hudson, 1995). The classic essay is Thomas H. Huxley, *Man's Place in Nature* (London: Williams and Norgate, 1863).

2. This passage on Neanderthal stereotypes draws on Gamble and Stringer, *Search*. Quote from 8.

3. Description in Stringer and Andrews, *Human Evolution*, 154–57 (see chap. 1, n. 2). For skin pigmentation, see Carles Lalueza-Fox et al., "A Melanocortin 1 Receptor Allele Suggests Varying Pigmentation Among Neanderthals," *Science* 2007 10,1126.

4. A vivid portrait of midwinter in the Canadian High Arctic appears in Maxwell Moreau, *Prehistory of the Eastern Arctic* (New York: Academic Press, 1985).

5. The description of last-interglacial and last-glaciation climate that follows is largely drawn from Tjeerd H. van Andel, "Glacial Environments I: The Weichselian Climate in Europe Between the End of OIS-3 Interglacial and the Last Glacial Maximum," in Tjeerd H. van Andel and William Davies, eds, *Neanderthals and Modern Humans in the European Landscape During the Last Glaciation: Archaeological Results of the Stage 3 Project* (Cambridge: McDonald Institute for Archaeological Research, 2003), 9–20.

6. See, for example, Eric Barron et al., "Glacial Environments II: Reconstructing the Climate of Europe in the Last Glaciation," in *Neanderthals and Modern Humans*, 57–79. Chaps. 6 and 7 are also valuable in this context.

7. M. P. Richards et al., "Isotopic Dietary Analysis of a Neanderthal and Associated Fauna from the Site of Jonzac (Charente-Maritime, France)," *Journal of Human Evolution* 55, no. 1 (2008): 179–85. See also Hervé Bocherens et al., "Isotopic Evidence for Diet and Subsistence Pattern of the Saint-Césaire I Neanderthal: Review and Use of a Multi-Source Mixing Model," *Journal of Human Evolution* 49, no. 1 (2005): 71–87.

8. For a beautifully illustrated account of mammoths for a popular audience, Adrian Lister and Paul Bahn, *Mammoths: Giants of the Ice Age* (Berkeley: University of California Press, 2007).

9. The provocative Guthrie, *Nature*, offers the professional insights of a paleontologist on the late Ice Age bestiary.

Chapter 4: The Quiet People

1. A summary of Lewis Binford's researches will be found in his *In Pursuit of the Past*, rev. ed. (Berkeley: University of California Press, 2001).

2. Did the Neanderthals throw spears? A discussion, based on a very small sample of actual Neanderthal and Cro-Magnon bones, is in Jill A. Rhodes and Steven E. Churchill, "Throwing in the Middle and Upper Paleolithic: Inferences from an Analysis of Humeral Retroversion," *Journal of Human Evolution* 56, no. 1 (2008): 1–10. I am grateful to Professor John Shea for discussion on the issue of range.

3. Philip Chase. *The Hunters of Combe Grenal: Approaches to Middle Paleolithic Subsistence in Europe* (Oxford: British Archaeological Reports International Series 286, 1986).

4. Katharine Scott, "Two Hunting Episodes of Middle Palaeolithic Age at La Cotte Saint-Brelade, Jersey (Channel Islands)," *World Archaeology* 12, no. 2 (1980): 137–52. For a wider survey of Neanderthal hunting, see Paul Mellars, *The Neanderthal Legacy* (Princeton, NJ: Princeton University Press, 1996). See also Mary Stiner and Steven Kuhn, "What's a Mother to Do? A Hypothesis About the Division of Labor and Modern Human Origins," *Current Anthropology* 47, no. 6 (2006): 953–80.

5. C. Farizy et al., *Hommes et Bisons du Paléolithique Moyen à Mauran (Haute-Garonne)* (Paris: CNRS *Gallia Préhistoire*, supplément 30, 1994).

6. Books on stone tool technology abound, but few of them are first-rate. One of the best is Jean-Luc Piel-Desruisseaux, *Outils Préhistoriques*, 5th ed. (Paris: Dunod, 2007).

7. Alban Derfleur et al., "Neanderthal Cannibalism at Moula-Guercy, Ardèche, France," *Science* 286 (1999): 128–31. For El Sidrón, A. Rosas et al, "Paleobiology and Comparative Morphology of a Late Neanderthal Sample from El Sidrón, Asturias, Spain," *Proceedings of the National Academy of Sciences* 103 (2006): 19266–71.

8. Neanderthal burial has long been a controversial subject, although most scholars now accept that they interred some of their dead. For a review, see Robert H. Gargett, "Grave Shortcomings: The Evidence for Neanderthal Burial," *Current Anthropology* 30, no. 2 (1989): 157–90.

9. This passage draws on Steven Mithen, *The Singing Neanderthals* (Cambridge, MA: Harvard University Press, 2006). This beautifully argued and written book is controversial and stimulating, as is all Mithen's work.

10. For a survey, see Philip Lieberman, *Uniquely Human: The Evolution of Speech, Thought, and Selfless Behavior* (Cambridge, MA: Harvard University Press, 1991). Also see Philip Lieberman, "The Evolution of Human Speech," *Current Anthropology* 48, no. 1 (2007): 39–66.

11. Mithen, *Singing*, 228.

12. Steven Mithen, *The Prehistory of the Mind* (London: Thames and Hudson, 1996). This is a wonderfully provocative essay on early cognitive skills.

13. Frederick L. Coolidge and Thomas Wynn. *The Rise of Homo sapiens: The Evolution of Modern Thinking.* (New York: Wiley-Blackwell, 2009).

Chapter 5: The Ten Thousandth Grandmother

1. Björn Kurtén, *Singletusk: A Story of the Ice Age* (New York: Pantheon, 1986), 61. Kurtén, an eminent paleontologist, wrote several wonderful novels about the Ice Age, which reflect his profound knowledge of the Arctic and of Ice Age animals.

2. Dorothy Garrod (1892–1969) was the Disney Professor of Archaeology at Cambridge University (1938–1952), the first woman ever to hold a chair at Oxford or Cambridge. She is best known for her excavations at Mount Carmel but also worked at other sites in the Near East, on Gibraltar, and in Kurdistan. The Mount Carmel excavations are described in Dorothy Garrod and Dorothea Bate, *The Stone Age of Mt. Carmel* (Oxford: Clarendon Press, 1937).

3. Ted McCown and Arthur Keith, *The Stone Age of Mt. Carmel*, vol. 2, *The Fossil Human Remains from the Levalloiso-Mousterian* (Oxford: Clarendon Press, 1939).

4. It's best to look at the latest version of her argument rather than the original paper: Dorothy Garrod, "The Relations Between Southwest Asia and Europe in the Late Paleolithic Age with Special Reference to the Origins of the Upper Paleolithic Blade Cultures," *Journal of World History* 1 (1953): 13–37.

5. Rebecca Cann et al., "Mitochondrial DNA and Human Evolution," *Nature* 325 (1987): 31–36. Quote from 31.

6. A good summary of genetics and modern human origins for a general audience appears in Wells, *Journey* (see chap. 1, n. 5). The Ingman research mentioned in this box is published in Michael Ingman et al., "Mitochondrial Genome Variation and the Origin of Modern Humans," *Nature* 408 (2000): 708–13. For a recent summary of the controversial genetic evidence for the spread of modern humans, see Stephen Oppenheimer, "The Great Arc of Dispersal of Modern Humans: Africa to Australia," *Quaternary International* 30 (2008): 1–12.

7. The controversies are summarized in Wells, *Journey*, chaps. 2–4.

8. Summary in P. M. Vermeersch, "The Upper and Late Paleolithic of Northern and Eastern Africa," in F. Klees and R. Kuper, eds., *New Light on the Northern African Past* (Cologne, Germany: Heinrich Barth Institute, 1992), 99–154. Thermoluminescence dating measures the accumulated radiation dose, and hence the age, of crystalline minerals in heated objects such as those made of lava or ceramics, as well as geological sediments.

9. John J. Shea, "Neanderthals, Competition, and the Origin of Modern Human Behavior in the Levant," *Evolutionary Anthropology* 12, no. 4 (2003): 173–87. Also see Francesco d'Errico et al., "Archaeological Evidence for the Emergence of Language, Symbolism, and Music—an Alternative Multidisciplinary Perspective," *Journal of World Prehistory* 17, no. 1 (2003): 1–70.

10. The African evidence: For Omo Kibish, see John J. Shea et al., "Context and Chronology of Early *Homo Sapiens* Fossils from the Omo Kibish Formation, Ethiopia," in Paul Mellars et al., eds., *Rethinking the Human Revolution* (Cambridge: McDonald Institute for Archaeological Research, (2007)), 153–76. For Herto, see T. D. White et al., "Pleistocene *Homo Sapiens* from Middle Awash, Ethiopia," *Nature* 423 (2003): 742–47.

11. Wells, *Journey*, chap. 3.

12. Christopher A. Scholz et al., "East African Megadroughts Between 135 and 75 Thousand Years Ago and Bearing on Early-Modern Human Origins," *Proceedings of the National Academy of Sciences* 104 (2007): 16416–21.

13. The Mount Toba eruption almost beggars description in terms of its scale and impact on humanity. Only recently has the full extent of the devastation caused by the eruption come to wider notice, and many details of the magnitude of the disaster and its effects on humanity remain controversial. For an excellent summary with an extensive bibliography, see Michael R. Rampino and Stanley Ambrose, "Volcanic Winter in the Garden of Eden: The Toba Supereruption and the Late Pleistocene Human Population Crash," *Geological Society of America Special Paper 345* (2000).

14. This is well described in Henry and Elizabeth Stommel, *Volcano Weather: The Story of 1816, the Year Without a Summer* (Newport, RI: Seven Seas Press, 1983). Simon Winchester describes the Krakatau explosion in his *Krakatoa: The Day the World Exploded; August 27, 1883* (New York: Harper Perennial, 2005).

15. Professor Stanley Ambrose: personal communication. I am grateful to him for his assistance with this passage.

16. The literature on this subject is proliferating rapidly. A summary of the debates and major issues will be found in Sally McBrearty, "Down with the Revolution," Mellars et al., *Rethinking*, 133–152.

17. For Blombos, see Christopher Stuart Henshilwood, "Fully Symbolic *Sapiens* Behavior: Innovations in the Middle Stone Age at Blombos Cave, South Africa," Mellars et al., *Rethinking*, 123–32. For Sibudu, see Lucinda Backwell et al., "Middle Stone Age Bone Tools from the Howiesons Poort Layers, Sibudu Cave, South Africa," *Journal of Archaeological Science* 35, no. 6 (2007): 1566–80. A valuable series of papers, published after this book was written, appear in Marlize Lombard, Christine Sievers, and Valerie Wood, eds., *Current Themes in Middle Stone Age Research* (Vlaeberg, South Africa: South African Archaeological Society, 2008).

18. A summary of the raw material procurement issue appears in Stanley H. Ambrose, "Howiesons Poort Lithic Raw Material Procurement Patterns and the Evolution of Modern Human Behavior: A Response to Minichillo (2006)," *Journal of Human Evolution* 50 (2006): 365–69.

19. Richard Lee, *The !Kung San* (Cambridge: Cambridge University Press, 1979).

20. A good survey for the general reader is Klein and Edgar, *Dawn* (see chap. 1, n. 5).

Chapter 6: Great Mobility

1. A huge literature surrounds the issues in this and the following paragraphs. Excellent bibliographies and summaries will be found in Anna Belfer-Cohen and A. Nigel Goring-Morris, "From the Beginning: Levantine Upper Palaeolithic Cultural Change and Continuity," Mellars et al., *Rethinking*, 199–206 (see chap. 5, n. 10). See also John J. Shea, "Transitions or Turnovers? Climatically-Forced Extinctions of *Homo Sapiens* and Neanderthals in the East Mediterranean Levant," *Quaternary Science Reviews* 30 (2008): 1–18.

2. For a summary and references, see Shea, ibid., 7–9.

3. John J. Shea, "The Origins of Lithic Projectile Point Technology: Evidence from Africa, the Levant, and Europe," *Journal of Archaeological Science* 33 (2006): 823–46. The origin of spear throwers is still largely a matter of intelligent speculation. They are a weapon that is particularly useful in open terrain. It is entirely possible that they originated in tropical Africa, and even more likely that modern humans in the Levant used them. In contrast, bows and arrows are more useful in wooded landscapes, for they can be used to shoot at all angles. I am grateful to Professor Shea for stimulating discussion on this complex topic. For throwing abilities, see Rhodes and Churchill, "Throwing" (see chap. 4, n. 2).

4. Anyone venturing into these controversial academic waters faces an enormous, usually contradictory literature. The summaries mentioned here all contain wide-ranging and useful bibliographies. See Ofer Bar-Yosef, "The Dispersal of Modern Humans in Eurasia: A Cultural Interpretation," Mellars et al., *Rethinking*, 207–18. A list of this author's many contributions to the subject will be found there. See also Paul Mellars, "The Impossible Coincidence: A Single-Species Model for the Origins of Modern Human Behavior in Europe," *Evolutionary Anthropology* 14 (2005): 12–27.

5. The Campanian eruption is a new player on the Cro-Magnon stage. See B. Giaccio et al., "The Campanian Ignimbrite (c. 40ka BP) and Its Relevance for the Timing of the Middle to Upper Palaeolithic Shift: Timescales and Regional Correlations," in Nicholas J. Conard, ed., *When Neanderthals and Modern Humans Met* (Tübingen, Germany: Kerns Verlag, 2006), 89–97. See also following note.

6. John Hoffecker et al., "From the Bay of Naples to the River Don: The Campanian Ignimbrite Eruption and the Middle to Upper Paleolithic Transition

in Eastern Europe," *Journal of Human Evolution* 30 (2008): 1–13. See also the same author's *Desolate Landscapes: Ice Age Settlement in Eastern Europe* (New Brunswick, NJ: Rutgers University Press, 2002), chap. 5. Paleomagnetic dating is based on changes in the orientation and intensity of the earth's magnetic field that have occurred over time. In archaeomagnetic dating, oriented specimens are recovered from baked immobile archaeological features, such as the soil surrounding a hearth, in order to determine the direction of the geomagnetic field at the time they were formed. Such dates have to be calibrated with radiocarbon and other readings.

7. João Zilhão et al., "The Peştera cu Oase People, Europe's Earliest Modern Humans," Mellars et al., *Rethinking*, 249–63.

8. Erik Trinkaus et al., "The Peştera cu Oase and Early Modern Humans in Southeastern Europe," Conard, *When Neanderthals*, 145–64.

9. The chronology of the first modern-human settlement of Europe is fraught with uncertainties, many of them caused by difficulties with the radiocarbon calibration. The chronologies given in these pages for the period of first settlement are the best currently available, but they are in a state of flux. Here are some papers that discuss the issues, often in considerable technical detail: S. P. E. Blockley et al., "The Middle to Upper Paleolithic Transition: Dating, Stratigraphy, and Isochronous Markers," *Journal of Human Evolution* 55 (2008): 764–71; Paul Mellars, "A New Radiocarbon Revolution and the Dispersal of Modern Humans in Eurasia," *Nature* 439 (2006): 932–935.

10. I am grateful to Dr. John Hoffecker for drawing my attention to new research on the Kostenki-Borschchevo kill sites. John F. Hoffecker. "The Spread of Modern Humans in Europe," *Proceedings of the National Academy of Sciences* 10.1073 (2009).

11. Henri Breuil, "Les Subdivisions de Paléolithique Supérieur et Leurs Signification," *Congrés Internationale d'Anthropologie et d'Archéologie Préhistorique*, Geneva (1912).

12. Brooke S. Blades, *Aurignacian Lithic Technology* (New York: Kluwer Academic, 2001).

13. Garrod, "Relations" (see chap. 5, n. 4).

14. Paul Mellars, "Archaeology and the Dispersal of Modern Humans in Europe: Deconstructing the Aurignacian," *Evolutionary Anthropology* 15 (2006): 167–82.

15. Bernhard Weninger and Olaf Jöris, "A 14C Calibration Curve for the Past 60 Ka: The Greenland-Hulu U/Th Timescale and Its Impact on Understanding the Middle to Upper Paleolithic Transition in Western Eurasia," *Journal of Human Evolution* 55 (2008): 772–81. Uranium-thorium dating is a radiometric dating

technique commonly used to determine the age of carbonate materials such as stalagmites or coral. It calculates the age of a sample from the degree to which equilibrium has been restored between the radioactive isotope thorium-230 and its radioactive parent uranium-234.

16. Mellars, "New Radiocarbon," 773ff.

17. The controversies over the Châtelperron issue rage unabated and are of interest mainly to specialists. For a critical summary of the different viewpoints, see Paul Mellars et al., "Confirmation of Neanderthal/Modern Human Interstratification at the Chatelperronian Type-Site," *Proceedings of the National Academy of Sciences*, 104 no. 3 (2007): 3657–62. My discussion is based on this paper and on Mellars, "New Radiocarbon." Another major, and opposing, analysis is in João Zilhão and Francesco d'Errico, "The Chronology and Taphonomy of the Earliest Aurignacian and Its Implications for the Understanding of Neanderthal Extinction," *Journal of World Prehistory* 13, no. 1 (1999): 1–68.

18. For an extended discussion of the various theories surrounding Neanderthal extinction, see Conard, *When Neanderthals*, 145–64.

19. Reported in *Observer* (UK), May 17, 2009. The original reference was unavailable at the time of writing.

20. I am grateful to Dr. Chris Stringer for discussion on the dating of Neanderthal extinction in Spain, which, given new calibration formulas, is considerably earlier than the twenty-eight thousand to twenty-four thousand years ago sometimes quoted in the literature.

Chapter 7: The Realm of the Lion Man

1. For an admirable summary of Australian Aboriginal history, see Peter Hiscock, *Archaeology of Ancient Australia* (Abingdon, UK: Routledge, 2008).

2. J. Hahn, "La Statuette Masculine de la Grotte de Hohlenstein-Stadel (Wurttemberg)," *L'Anthropologie* 754 (1971): 233–43.

3. J. Hahn, "Aurignacian Art in Central Europe," in Heidi Knecht et al., eds., *Before Lascaux: The Complex Record of the Early Upper Palaeolithic* (Boca Raton, FL: CRC Press, 1993), 229–57.

4. Nicholas J. Conard, "A Female Figurine from the Basal Aurignacian of Hohle Fels Cave in Southwestern Germany," *Nature* 459, no. 7244 (2009): 248–52.

5. Mithen, *Singing Neanderthals*, chaps. 15 and 16 (see chap. 4, n. 9).

6. An extended discussion of shamanism appears in David Lewis-Williams, *The Mind in the Cave* (London: Thames and Hudson, 2002). Anyone doubtful about the power of shamans and altered states of consciousness in human societies should read Lawrence Sullivan, *Icanchu's Drum* (New York: Free Press, 1989).

7.   Shamanism and altered states of consciousness in the context of the Cro-Magnons, and especially their rock art, are controversial topics. For a discussion of the subject, see Lewis-Williams, *Mind*, and Jean Clottes and David Lewis-Williams, *Shamans of Prehistory* (New York: Harry N. Abrams, 1998). Opponents of shamanism publish mainly in academic journals; consult a specialist for references.

8.   J-M. Chauvet, E. B. Deschamps, and C. Hillaire, *Chauvet Cave: The Discovery of the World's Oldest Paintings* (London: Thames and Hudson, 1996). Quote from 42. Also see Jean Clottes, *Chauvet Cave: The Art of Earliest Times* (Salt Lake City: University of Utah Press, 2003).

9.   Paul Bahn and Paul Pettit, "Art and the Middle-to-Upper Paleolithic Transition in Europe: Comments on the Archaeological Arguments for an Early Upper Paleolithic Antiquity of the Grotte de Chauvet Art," *Journal of Human Evolution* 55, no. 5 (2008): 908–17.

10.   William Shakespeare quotes. Owl: From the poem *Venus and Adonis*, line 552; Fatal bellman: *Macbeth*, Act II, scene 2, line 50.

11.   Jean Clottes, *Cave Art* (London: Phaidon, 2008), 50.

12.   Again, the literature is diffuse and enormous. For a summary, see Paul G. Bahn and Jean Vertut, *Images of the Ice Age* (Berkeley: University of California Press, 1997), chap. 11. See also Lewis-Williams, *Mind*, chap. 2. For rock art on the global stage, see David S. Whitley, ed., *A Handbook of Rock Art Research* (Walnut Creek, CA: AltaMira Press, 2001).

13.   David Lewis-Williams, *Believing and Seeing: Symbolic Meanings in Southern San Rock Paintings* (London: Academic Press, 1981), and David Lewis-Williams and Thomas Dowson, *Images of Power: Understanding South African Rock Art*, 2nd ed. (Cape Town: C. Struik, 1999).

14.   Hiscock, *Archaeology*, chap. 11.

Chapter 8: Fat, Flints, and Furs

1.   This passage is based on historic arctic tailoring practices. See Cornelius Osgood, *Ingalik Material Culture* (New Haven, CT: Human Relations Area Files, 1970), and Richard Nelson, *Hunters of the Northern Ice* (Chicago: University of Chicago Press, 1969).

2.   For trapping, see Osgood, *Ingalik*, 336ff.

3.   Van Andel and Davies, *Neanderthals*, chaps. 2, 5–7 (see chap. 3, n. 5).

4.   Will Roebroeks and Clive Gamble, eds., *The Middle Palaeolithic Occupation of Europe* (Leiden, Netherlands: University of Leiden Press, 1999) See also the articles in Conard, *When Neanderthals* (see chap. 6, n. 5).

5.   Hoffecker, *Desolate Landscapes*, chaps. 5 and 6 (see chap. 6, n. 6).

6.   Figures and discussion based on Guthrie, *Nature*, 337–60 (see chap. 2, n. 1).

7.   I introduced the notion of the Swiss Army Knife's relationship to Cro-Magnon technology in my *Journey from Eden*, (London: Thames & Hudson 1990) (154–58), but the real credit for the idea should go to Shelly Lowenkopf, who put the idea into my head. For blade technology generally, see Piel-Desruisseaux, *Outils Préhistoriques*, pt. 1 (see chap. 4, n. 6).

8.· Ibid., 111ff.

9.   André Rigaud, "La Technologie du Burin Appliquée au Material Osseux de la Garenne (Indre)," *Bulletin de la Société Préhistorique Française* 69, n. 4 (1972): 104–08. For bone-working description, see Osgood, *Ingalik*, 301.

10.   Nelson, *Hunters*, 176. He refers to caribou-skin bags, but reindeer bags would have been similar.

11.   This passage relies heavily on Nelson, *Hunters*, 249–60, which is a definitive account of northern clothing.

12.   Ibid., 318–19. See also Osgood, *Ingalik*, 265–66.

13.   My source for this section was Nelson, *Hunters*, chap. 18.

Chapter 9: The Gravettians

1.   For ptarmigan traps, see Cornelius Osgood, *Ingalik Social Culture* (New Haven, CT: Yale University Press, 1958), 336ff.

2.   A synthesis of this region appears in Jiri Svoboda et al., *Hunters Between East and West: The Paleolithic of Moravia* (New York: Plenum Press, 1996).

3.   A summary of the Gravettian appears in Clive Gamble, *The Palaeolithic Societies of Europe* (Cambridge: Cambridge University Press, 1999), 287ff.

4.   M. Oliva, "A Gravettian site with Mammoth-Bone Dwelling in Milovice (Southern Moravia)," *Anthropologie* 27 (1988): 265–71.

5.   Lister and Bahn, *Mammoths* (see chap. 3, n. 8).

6.   Quoted in Ibid., 52.

7.   N. K. Vereshchagin, "The Berelekh 'Cemetery' of Mammoths," *Proceedings of the Zoological Institute, Leningrad* 72 (1977): 3–50. This article is in Russian. See also Lister and Bahn, *Mammoths*, 62–63, 196–70.

8.· Everyone should read Elizabeth Marshall Thomas's *Reindeer Moon* (New York: Pocket Books, 1991), which gives a vivid impression of the difficulties and tensions of hunter-gatherer life. Set in the late Ice Age, the novel is based in part on Marshall's experiences among the San of the Kalahari Desert.

9.   For a summary description, see Svoboda et al., *Hunters*, 223–36.

10. For a summary of this important site, ibid., 209–12. For the triple grave, see Kurt W. Alt et al., "Twenty-Five Thousand Year-Old Triple Burial from Dolní Věstonice: An Ice Age Family?," *American Journal of Physical Anthropology* 102 (1997): 123–31.

11. The source for this section is Hoffecker, *Desolate Landscapes*, chaps. 5 and 6 (see chap. 6, no. 6).

12. For the definitive account of Mezhirich, see Olga Soffer, *The Upper Palaeolithic of the Central Russian Plain* (New York: Academic Press, 1985).

13. For trapping analogies, see Osgood, *Ingalik Social Culture*, 196ff.

14. Francis B. Harrold, "Variability and Function Among Gravette Points from Southwestern France," in Gail Larsen Peterkin, Harvey M. Bricker, and Paul Mellars, eds., *Hunting and Animal Exploitation in the Later Palaeolithic and Mesolithic of Eurasia* (Washington, DC: Papers of the American Anthropological Association 4, 1993), 69–83.

15. Heidi Knecht, "Early Upper Paleolithic Approaches to Bone and Antler Projectile Technology," Peterkin, Bricker, and Mellars, *Hunting* 33–48.

16. Peter Savolainen et al., "Genetic Evidence for an East Asian Origin of Domestic Dogs," *Science* 298 (2002): 1613–1620. For Eliseevichi I dogs, see Mikhail V. Sablin and Gennady A. Khlopachev, "The Earliest Ice Age Dogs: Evidence from Eliseevichi I," *Current Anthropology* 43, no. 5 (2002): 795–99.

17. For a summary account of the Sungir site and burials with Russian references, see Hoffecker, *Desolate Landscapes*, 151ff. I also drew on Randall White, *Prehistoric Art* (New York: Harry N. Abrams, 2003), 145–48.

18. A huge literature surrounds Venus figurines, much of it at best speculative. For an especially intelligent summary, see Bahn and Vertut, *Images*, 98ff (see chap. 7, n. 12).

19. Guthrie, *Nature*, chap. 6 (see chap. 2, n. 1), presents a well-grounded, wide-ranging interpretation of female figurines and other depictions of women in Cro-Magnon art, which is both entertaining and based on solid science. As he says in the table of contents, "You won't be bored." He's right!

20. For an extended discussion, see Lee, *!Kung*, (see chap. 5, n. 19).

21. Guthrie, *Nature*, 371.

Chapter 10: The Power of the Hunt

1. Farley Mowat, *People of the Deer* (Boston: Little, Brown, 1952). I drew unashamedly on Mowat's classic description of caribou hunters in Canada's Barren Lands for this reconstruction and subsequent paragraphs. This remarkable, and beautifully written, book is a treasure house of information on northern life.

For late Ice Age hunting, see Mary C. Stiner, "Carnivory, Coevolution, and the Geographic Spread of the Genus *Homo*," *Journal of Archaeological Research* 10, no. 1 (2002): 1–63. For grease rendering by Cro-Magnons, see Mary C. Stiner, "Zooarchaeological Evidence for Resource Intensification in Algarve, Southern Portugal," *Promontoria* 1, no. 1 (2003): 1–58.

2. Bryan Gordon, *Of Men and Reindeer Herds in French Magdalenian Prehistory* (Oxford: British Archaeological Reports International Series 390, 1988).

3. The short visitor's guide to Abri Pataud is an admirable summary for laypeople: Brigitte and Giles Delluc, *Visiter l'abri Pataud* (Bordeaux, France: Éditions Sud-Ouest, 1998). There are numerous technical publications. For a good starting point, see Hallam L. Movius Jr., ed., *Excavation of the Abri Pataud, Les Eyzies (Dordogne)* (Cambridge, MA: Peabody Museum of Archaeology and Ethnology, Harvard University, 1975).

4. Guthrie, *Nature*, chap. 2, covers this subject in detail and was the source for these paragraphs (see chap. 2, n. 1).

5. For a useful summary guide to Cro-Magnon cave-art sites you can visit, see Paul G. Bahn, *Cave Art: A Guide to the Decorated Ice Age Caves of Europe* (London: Francis Lincoln, 2007).

6. Guthrie, *Nature*, 61.

7. Paul Bahn and Jean Vertut, *Journey Through the Ice Age* (Berkeley, CA: University of California Press, 1997), 111–12. For the Roc de Sers ibex, see Clottes, *Chauvet Cave Art*, 222–23 (see chap. 7, n. 1).

8. For a brief description of Pech Merle, see Bahn, *Cave Art*, 96–101.

9. For an overview on Gargas, see Bahn, *Cave Art*, 112–19. See also Marc Groenen, "Les Representations de Mains Negatives dans les Grottes de Gargas et de Tibiran (Hautes-Pyrénées)," *Bulletin de la Société Royale Belge d'Anthropologie et de Préhistoire* 99 (1988): 81–113.

10. See Hoffecker et al., "Evidence" (see chap. 6, n. 10).

11. For a useful summary of horses and their history, see Sandra L. Olsen, ed., *Horses Through Time* (Boulder, CO: Roberts Rinehart for the Carnegie Museum of Natural History, 2003).

12. My account of Solutré is based on Jean Combier and Anta Montet-White, eds., *Solutré 1968–1998* (Paris: Société Préhistorique Française, Memoir 30, 2002).

Chapter 11: The Magdalenians

1. For the definitive account, see Norbert Aujoulet, *Lascaux: Movement, Space, and Time* (New York: Harry N. Abrams, 2005).

2. For a detailed account, see Combier and Montet-White, *Solutré* (see chap. 10, n. 12).

3. Piel-Desruisseaux, *Outils Préhistoriques*, 45, 130 (see chap. 4, n. 6).

4. Breuil, "Subdivisions" (see chap. 6, n. 11).

5. André Leroi-Gourhan and M. Brézillion, *Fouilles de Pincevent: Essai d'Analyse Ethnographique d'un Habitat Magdalénien* (Paris: CNRS *Gallia Préhistoire*, supplément 7, 1972).

6. Tanning reindeer, or rather caribou, hides proceeded in similar ways throughout the historic North, albeit with minor local variations. This passage is based on Osgood, *Ingalik Social Culture* (see chap. 9, n. 11) 200.

7. Henri Laville et al., *Rockshelters of the Périgord* (New York: Academic Press, 1980).

8. Some of this reconstruction is based on information in Osgood, *Ingalik Social Culture* (see chap. 9, n. 1), 342ff.

9. For Magdalenian technology, see Piel-Desruisseaux, *Outils Préhistoriques*, pts. 2 and 3.

10. Cap Blanc is well summarized in Bahn, *Cave Art*, 54–57 (see chap. 10, n. 5).

11. Descriptions of portable Magdalenian art abound. A good one appears in Bahn and Vertut, *Journey*, chap. 7 (see chap. 10, n. 7).

12. Henri Bégouin and Henri Breuil, *Les Cavernes du Volp: Trois-Frères, Tuc d'Audoubert, à Montesquieu-Avantés (Ariège)* (Paris: Arts et Métiers Graphiques, 1958).

13. Antonio Beltrán et al., *Altamira* (Paris: Le Seuil, 1998).

14. The best guide for visitors is Bahn, *Cave Art*, 137–47.

Chapter 12: The Challenge of Warming

1. Jean Clottes, *Les Cavernes de Niaux* (Paris: Le Seuil, 1995). Clottes's interpretation appears in his *Cave Art*, 194 (see chap. 7, n. 11). The original entrance no longer exists, replaced by a large one blasted from the hillside.

2. Niaux is one of the few caves where extensive pigment and radiocarbon studies have been made. Jean Clottes, "Paint Analyses from Several Magdalenian Caves in the Ariège Region of France," *Journal of Archaeological Science* 20 (1993): 223–35.

3. Neil Roberts, *The Holocene: An Environmental History*, 2nd ed. (New York: Wiley/Blackwell, 1998).

4. Steven Mithen, *After the Ice Age: A Global Human History 20,000–5000 BC* (Cambridge, MA: Harvard University Press, 2003). This fascinating, closely argued book was an important source for this chapter.

5. L. G. Straus and G. A. Clark, *La Riera Cave: Stone Age Hunter-Gatherer Adaptations in Northern Spain* (Tempe, AZ: Anthropological Research Papers of Arizona State University, 17, 1986).

6. Claude Courand, *L'Art Azilien: Origine—Survivance* (Paris: CNRS *Gallia Préhistoire*, supplément 20, 1985).

7. El Mirón is summarized in Lawrence Guy Straus, "Last Glacial Settlement in Eastern Cantabria (Northern Spain)," *Journal of Archaeological Science* 29 (2002): 1403–14.

8. R. A. Housley et al., "Radiocarbon Evidence for the Late glacial Recolonization of Northern Europe," *Proceedings of the Prehistoric Society* 63 (1997): 25–54.

9. A. Leroi-Gourhan and M. Brézillion, *Fouilles de Pincevent: Essai d'Analyse Ethnographique d'un Habitat Magdalénien* (Paris: CNRS *Gallia Préhistoire*, supplément 7, 1972). See also J. G. Enloe et al., "Patterns of Faunal Processing at Section 27 of Pincevent: The Use of Spatial Analysis and Ethnoarchaeology in the Interpretation of Archaeological Site Structure," *Journal of Anthropological Archaeology* 13 (1994): 105–24.

10. Nicole Pigeot, "Technical and Social Actors: Flint Knapping Specialists and Apprentices at Magdalenian Etiolles," *Archaeological Review from Cambridge* 9 (1990): 126–41.

11. The best source on Meiendorf, with a comprehensive bibliography, is Bodil Bratlund, "Hunting Strategies in the Late Glacial of Northern Europe: A Survey of the Faunal Evidence," *Journal of World Prehistory* 10, no. 1 (1996): 1–48.

12. G. Bosinski and G. Fisher, *Die Menschendarstellungen von Gönnersdorf der Ausgrabung von 1968* (Wiesbaden, Germany: Steiner, 1974). Bosinski describes the depictions of women in "The Representation of Female Figures in the Rhineland Magdalenian," *Proceedings of the Prehistoric Society* 57 (1991): 51–64.

13. The most accessible description of Stellmoor with references is in Bratlund, "Hunting Strategies," 17ff.

14. The sunken North Sea landscape is often called Doggerland, after the submerged Dogger Bank, a notable modern-day fishing ground. Star Carr was originally described in J. G. D. Clark's classic monograph, *Star Carr* (Cambridge: Cambridge University Press, 1954). The latest update is Paul Mellars and Petra Dark, *Star Carr in Context* (Cambridge: McDonald Institute, and the Vale of Pickering Research Trust, 1999).

15. Summaries of the complex archaeology of this region will be found in Geoff Bailey and Penny Spikins, eds., *Mesolithic Europe* (Cambridge: Cambridge University Press, 2008), chaps. 1–4.

16. Saxton Pope, *A Study of Bows and Arrows* (Berkeley: University of California Press, 1923).

17. For Lepenski Vir, see D. Srejovic, *Europe's First Monumental Sculpture: New Discoveries at Lepenski Vir* (London: Thames and Hudson, 1989). For updates on chronology and other details, see Bailey and Spikins, *Mesolithic Europe,* chap. 10.

18. For a comprehensive summary of the origins of food production, see Graeme Barker, *The Agricultural Revolution in Prehistory* (New York: Oxford University Press, 2009).

19. A description of the LBK people appears in Barker, *Agricultural Revolution,* chap. 9.

20. For an excellent summary and critical analysis of this development, see J. Zilhão, "The Spread of Agro-pastoral Economies Across Mediterranean Europe," *Journal of Mediterranean Archaeology* 6 (1993): 5–63.

21. A discussion appears in Barry Cunliffe, *Europe Between the Oceans* (London and New Haven, CT: Yale University Press, 2008), chap. 3.

22. The literature is complex. See L. Luca Cavalli-Sforza and E. Minch, *The History and Geography of Human Genes* (Princeton, NJ: Princeton University Press, 1994), and M. R. Richards et al., Palaeolithic and Neolithic Lineages in the European Mitochondrial Gene Pool," *American Journal of Human Genetics* 59 (1996): 185–203. Also see Brian Sykes, "The Molecular Genetics of European Ancestry," *Philosophical Transactions of the Royal Society of London B.* 354 (1999): 131–39, and his *The Seven Daughters of Eve* (London: Transworld Publishers, 2001). A useful summary appears in Roger Lewin, "Ancestral Echoes," *New Scientist* 1089 (1997): 32–37.

Index

A Note on the Author

Brian Fagan is emeritus professor of anthropology at the University of California–Santa Barbara. Born in England, he did fieldwork in Africa and has written about North American and world archaeology and many other topics. His books on the interaction of climate and human society have established him as the leading authority on the subject; he lectures frequently around the world. He is the editor of *The Oxford Companion to Archaeology* and the author of *The Great Warming*; *Fish on Friday: Feasting, Fasting, and the Discovery of the New World*; *The Little Ice Age*; and *The Long Summer*, among many other titles.